GALANA

GALANA

Elephant, Game Domestication, and Cattle on a Kenya Ranch

MARTIN ANDERSON

Kim Eggers   June 1, 2014

Dear Kim
Galana was a great
adventure.
Martin Anderson

STANFORD GENERAL BOOKS
An Imprint of Stanford University Press
Stanford, California

Stanford University Press
Stanford, California

Printed in the United States of America on acid-free, archival-quality paper

Library of Congress Cataloging-in-Publication Data
Anderson, Martin, 1923- author.
  Galana : elephant, game domestication, and cattle on a Kenya ranch / Martin Anderson.
    pages cm
  ISBN 978-0-8047-8924-0 (cloth : alk. paper)
  1. Galana Ranch--History. 2. Ranching--Kenya--History--20th century. 3. Big game ranching--Kenya--History--20th century. 4. Elephants--Kenya--History--20th century.
  5. Poaching--Kenya--History--20th century. 6. Wildlife management--Kenya--History--20th century. I. Title.
  SF196.K4A44 2013
  636.2'01096762--dc23
                          2013005413
Typeset by Bruce Lundquist in 10/14 Minion

Oh, I realize that we civilize,
And the work that we do is fine,
When we lay the trails for the
gleaming rails
Of a pioneering line.
But soon they'll push, till there's no
more bush,
And never a bushman's shrine,
And when the day's come where will be
the home
For a soul that is made like mine?

Brian Brooke

CONTENTS

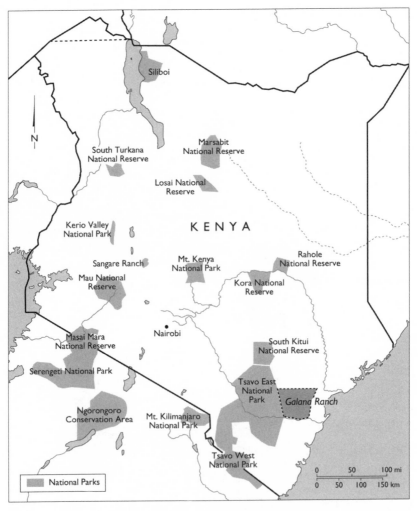

Map of Kenya showing Galana Ranch (approximately 1% of Kenya).

FOREWORD

Dr. Dame Daphne Sheldrick DBE MBE MBS DVMS

There is a saying that "To Hear is to Know; to See is to Believe, and to Do is to Understand". Martin Anderson, essentially professional and entrepreneurial in all he undertakes, knows this all too well, having worked in Africa, close to Nature and all animals, both domestic and wild, but also having to deal with the realities of corruption, cronyism, and envy. Yet, throughout his Galana experiment, Martin never crossed the boundaries of honesty, fairness, and integrity. The Galana experiment was, in fact, a blue print of how to manage marginal land sustainably, efficiently, and profitably, and, but for bureaucratic human failings, could have been the example for others to follow. Adventurous and brave, Martin proved himself an expert rancher and leader of men as well as a caring conservationist sensitive to the natural environment and its wild inhabitants. His 1.6 million acre Galana Ranch was a model worthy of celebration that should have been welcomed and nurtured by officialdom rather than envied and rejected.

Said Confucius—"If you would foretell the future, study the Past". Now in the twilight years of our lives, and with the benefit of hindsight, Martin and I both understand this ancient saying to be so true. The problems that beset Galana Ranch during Martin's tenure were similar to those in neighbouring Tsavo National Park, the only difference being that our "cattle" were the iconic elephant herds of the day whom we fiercely protected against proposed "culling" for reasons that are obvious to all who know the politics and history of Kenya's elephants and their ivory tusks. Today, the poaching problem is still with us, and perhaps even more serious than it was then, because the elephants that remain are fewer. Added to the Somali incursion is that of the Chinese, now present throughout every Elephant Range State in Africa, and whose populous

---

Daphne Sheldrick's book on her experience in Africa is *An African Love Story: Love, Life and Elephants*. New York: Penguin, an imprint of Viking, 2012.

and wealthy homeland fuels the demand for ivory. Yet, in a country like Kenya where the sun shines every day and where there are still magic wild corners to enjoy where the natural world remains intact, there *is* hope—the hope that Martin's Galana example will one day be emulated by a more enlightened generation of young Kenyans in the corridors of power, who will nurture, protect, and care for the irreplaceable natural legacy with which their country has been blessed and which is the envy of the world. This hope is reflected in the Kenyans who diligently care for the elephant orphans of poaching now under the care of The David Sheldrick Wildlife Trust, established in memory of my late husband, and who do so with dedication and a sincere empathy and love, shedding tears at the death of every orphan that succumbs to the trauma of losing their beloved Elephant Family.

Martin and Illie Anderson were more than just neigbours to us in Tsavo, with whom we shared many magic moments, as well as many of the same headaches and problems. They were our closest friends whose counsel was always understanding, immensely wise, and, above all, genuinely compassionate and sincere. Like us, they worked out of a sense of dedication and in a way they knew was right. And they were there for me in my darkest hour when I lost my husband very suddenly at a young age and did not know how I would be able to cope. They quietly and unobtrusively ensured that my 13-year-old daughter, Angela, could return from school in South Africa to be with me during the holidays, and also ensured that her innate artistic talent could be further honed at Capetown University when she graduated from high school. Today, Angela runs The David Sheldrick Wildlife Trust and I know her father would be immensely proud of her achievements, but seldom a day passes that I am not endlessly grateful for the enduring and special friendship of Martin and Illie Anderson just when it was needed most. I know how deeply Martin must miss Illie, just as I still miss David, and I admire Martin for bravely penning such an accurate account of the unpredictable and insurmountable hurdles that cost him his African dream.

PREFACE

HUNDREDS OF ELEPHANTS, silhouetted by the sinking sun, drifted, dream-like, over the flat, featureless plains of eastern Kenya. The herds were moving, in a vast, timeless procession, over the edge of the horizon. It was the great gathering of the elephant clans at the beginning of the October rainy season.

From our Land Rover, 600 yards away, my wife, Illie, and I watched in near disbelief as the column of tuskers paraded majestically in front of us for half an hour. The herds were abandoning the stripped, overbrowsed woodlands of Tsavo National Park. In search of richer feeding grounds, they were heading for the adjacent Galana Ranch—2,500 square miles of wild, raw Africa for which Illie and I had, recently and improbably, become responsible.

In that moment I realized, with humility, that we were undertaking some-thing much bigger and more profound than we understood.

It was October 1968—five years after Kenya had won independence and six-teen months after its government had granted us, and two associates, a forty-five-year lease on this vast, unpeopled, untamed African landscape. Inspired by the spirit of *harambee*—the optimistic call, meaning "Let's work together," of Kenya's first president, Jomo Kenyatta—we were hoping to make a contribu-tion to the young nation. Our goals were to succeed in a uniquely challenging business venture and to manage and protect the wildlife in this huge, harsh stretch of bush near the Indian Ocean.

I was an entrepreneurial attorney from Hawaii, in my early forties, and to me, Galana seemed removed from the civilized world. It was not "beautiful" in the normal sense of the word. Its topography was almost relentlessly flat, except for a few hillocks, and its palette was unvaried, except for exquisite gradations of clay, burnt ochre, and rust, with flashes of green and silver. In this unchang-ing, monochromatic landscape, time and civilization did not exist—especially in the lingering half-light before dark, when night prowlers and predators began to wake.

We were idealistic and enthusiastic about what we might achieve and the life we could lead on these Kenyan grasslands. When we first arrived, Galana was a romantic relief from the modern world. Its single airstrip had been built in the 1920s by the legendary British hunter-explorer Denys Finch Hatton—the lover of Beryl Markham, author of *West with the Night*, and Karen Blixen, who wrote *Out of Africa* under the name Isak Dinesen. Galana's history, the immensity of its land—one percent of Kenya—and the vastness of its wild animal population made us conscious nearly every moment that our time there, legally and metaphysically, would be very brief. The beginnings and end of Galana's true calendar were as infinite and distant as the blazing stars.

Our obligation, under terms of the lease, was to ranch cattle on this grassland for the first time. That was no simple task, since Galana was dry and full of tsetse flies and ticks carrying bovine diseases. Our lease also required us to manage Galana's natural game population, help it grow to a level the land could support, and protect it against threatening imbalances. Beyond that, our dream was to use Galana's wildlife as a renewable resource to provide protein for Kenya's expanding population. In the 1960s, biologists, ecologists, and conservationists around the world were arguing that on marginal lands game ranching could be ecologically beneficial and economically more profitable than cattle ranching. Game ranches, however, had been developed on a major scale only in southern African nations, below the Zambesi River. Kenyan efforts had been very limited. I wanted to experiment with game domestication on a greater scale, with greater scientific discipline, than had ever been attempted in Kenya. We called our venture Galana Game and Ranching, reflecting our true priority—game before ranching.

Over the next two decades, we achieved those goals. We developed Galana into the largest cattle ranch in Kenya and perhaps all of Africa. Our 26,000 head included a government-registered stud herd of 400 breeding, humpbacked, pale-hided Boran cows, whose ancestors had arrived from north of Suez centuries ago. We learned how to introduce cattle into an area with virulent tsetse- and tick-borne disease, adapt our animals to the land, and breed resistant stock. We conducted scientific research for the domestication of wildlife, and we discovered that oryx were best suited to domestication. We employed 400 Kenyans, from a variety of tribes, to herd cattle, oryx, eland, buffalo, and antelope. We also developed miles of roads, airstrips, buildings, boreholes, pipelines, and dams and poured hundreds of thousands of dollars into their construction. But as soon as we entered the picture in 1967, we were also contending with politically complex, competing ambitions and ideas about wildlife management and conservation.

At the time, the biggest threat to Galana's habitat was the elephant population explosion in Tsavo National Park, a nearly 8,000-square-mile wildlife sanctuary. Elephants were thriving in Tsavo, and by 1966, their numbers had climbed to about fifteen thousand. There were so many elephants that their grazing was destroying the park's woodlands and threatening the wildlife they supported. Kenya's Game Department, at first, believed that cropping was the best way to control the elephant population. The government wanted us to shoot and kill up to three hundred elephant a year to reduce the size of the herds that moved between Galana and Tsavo.

We wanted nothing to do with the wholesale killing of wildlife.[1] At the Galana Ranch, we permitted hunters, in accordance with our lease terms, to shoot a few mature bulls a year for a large license fee—part of which went to the government and part of which helped pay for the protection and management of Galana's wildlife. Our hunting memberships discouraged a shoot-anything-that-moves mentality and ensured that only elephants of a minimum size, past breeding age, would be shot on the ranch.

As the price of ivory climbed drastically in the 1970s, however, poaching became a dangerous, disastrous problem. By 1971 gangs of poachers, armed with automatic weapons, had moved into Tsavo. With growing government

1. The "Letter of Allotment, Game Ranching Scheme," dated June 7, 1967, put us in possession of the Galana territory. In paragraph 16 of the "Special Conditions" section, the Kenya government stated:

Game cropping shall be carried out by the Grantee in accordance with the following rules: . . . (b) Game cropping shall be in accordance with an annual quota allotted by the Game Department and the Grantee shall rigidly adhere to the terms of the allotment of the quote. Both maximum and minimum quotas will be stipulate . . . (d) the Grantee shall pay to the Government of Kenya K.Shs. 100/= [a hundred shillings, equal at the time to five dollars] in respect of each elephant killed in the process of game cropping.

Shortly after receipt of the Letter of Allotment, Mike Prettejohn and I had a conference with the Honorable Bruce Mackenzie, then minister of agriculture in President Kenyatta's cabinet. We pointed out to the minister that it was unlikely that any public health office would approve for human consumption meat from elephant that had been shot miles away from a processing plant. Anyway, such cropping would quite properly arouse public condemnation. The minister agreed and excused performance of that condition.

It was particularly galling when government officials later gave any credence to wild rumors that Galana was killing elephant for sale in the ivory trade. Under the Letter of Allotment, we could have shot elephant for ivory and legally made hundreds of thousands of dollars. We refused that at the outset.

corruption, elephant tusks were becoming vehicles for laundering money and taking it out of the country. By 1973, when the black market price of ivory jumped from $50 a kilo to $700 a kilo, a thousand elephants a month were being slaughtered in Kenya.

The problem worsened after May 1977, when the government imposed a total ban on hunting, ostensibly to preserve what remained of Kenya's once boundless wildlife. Animal lovers around the world cheered the move, believing that something was at last being done to stop the eradication of Kenya's great game herds. Editorialists from New York to Tokyo praised President Kenyatta for a courageous decision to spare the nation's unique wildlife for the future.

The ban, of course, meant the immediate end of our hunting safaris and a loss of income. But it also removed the watchful eyes of law-abiding hunters and their clients from Kenya's game lands. Without the presence of professional hunters, trackers, skinners, and clients in the field to report illegal activities, poachers were free to continue their destruction of Kenya's wildlife. There was little cause for optimism, given the rumors that high-ranking officials and members of Kenyatta's family were involved in the illegal sale and exportation of skins and ivory.

As we feared, poaching increased dramatically after the hunting ban. Galana's game manager was soon finding two, three, or four bloody elephant carcasses a week, their tusks hacked out of their heads. By 1987, poachers had slaughtered more than 5,000 of the 6,000 elephants counted at Galana in 1967. They spared fewer than 6,000 in Tsavo National Park. The elephant herds were being wiped out—and, ultimately, the Galana Ranch was too big to succeed.

As Kenya's population exploded, it became politically untenable for a few white ranchers to control so much territory in a country where land was un-available to many Africans who needed it. Our Galana experiment ended in 1989—less than halfway through the length of our lease. The elephant migra-tion across the horizon that Illie and I witnessed in our early years at Galana will likely never be seen again on such a massive scale. The cattle have vanished, and the roadways have disappeared. The dams and water holes are no longer tended, although the pipelines remain. Galana's carefully built lodge, homes, and other buildings have been razed or have fallen into disuse, and only a small tourist accommodation remains. Almost all of the ranching activity that once thrived at Galana has vanished.

Still, the experience was invaluable—to us and as a lesson in conservation, political realities, and Africa's profound challenges. Galana was an adventure, a way of life, but it was also an experiment in wildlife management, animal domestication, and scientific research in a dangerous, dynamic political landscape. Although Galana ultimately failed, it was a successful laboratory for new ideas and a case study in the interplay of hunting and game management policies.

In this book, I recount the story of how and why the Galana Ranch came into being, how it grew, and what it accomplished. I owe this record of the Galana experiment to the courageous and devoted men and women—American, English, Waliangulu, Masaai, Orma, Turkana, Kamba, and others—who were responsible for its contributions.

More, I owe it to Illie, whose daily journal of our time at Galana gave life to this book.

Martin Anderson (center), with Mike Prettejohn (left) and Gilfrid Powys (right), together in Gilfrid's ranch living room, July 2012.

ACKNOWLEDGMENTS

I STARTED THIS BOOK AT LEAST THREE TIMES as a short family history of what Illie and I did and the life that we led over the years when we were absent in Kenya. It would never have been more than some notes if not for the discipline of three skillful and very competent editors—Bruce Frankel, Alan Harvey, and Susan Wels—who enforced deadlines, made corrections, selected pictures, and generally reorganized text. They brought the book to its conclusion.

On reflection, I would not have even made a start if not for friends who encouraged me to set forth a bit of our limited experience in the transition of an African country from colony to independence. The real progress in expanding an adventure story and Illie's daily journal entries into what is now *Galana* is due to Dr. Michael Keller, Stanford University Librarian. He suggested that I focus the book more on the history of an entrepreneur during Kenya's move to democracy.

To make that shift, I relied on the scientific publications of Dr. Mark Stanley Price of Oxford University, Dr. John King of Kenya Wildlife Services, Dr. T. J. Wacher, and Brian Heath. In addition, the African Wildlife Leadership Foundation (AWLF), through its Kenya director, John Minot, as well as Texas A&M's Caesar Kleberg Wildlife Research Institute, directed by Chris Field, provided needed documentation of our domestication research.

I owe special thanks to those who shared their memories and expertise, including Dame Daphne Sheldrick, Ambassador William C. Harrop, Ambassador Elinor Greer Constable, and Kitty Hempstone. It goes without saying that I owe a great debt of gratitude to those who devoted important parts of their lives to the ranch and contributed their recollections to this book, including Tony Dyer, Chris Flatt, Barney Gasston, Jessica and Henry Henley, Jeff and Jennifer Lewis, Mark Stanley Price, Tony Seth-Smith, and Angela Sutton.

David W. Brady and William Meehan made special efforts to keep the book on track.

I am, of course, most grateful to Gilfrid Powys and Mike Prettejohn, without whom there never would have been a Galana Ranch.

GALANA

CHAPTER ONE

IMAGINING AFRICA

RANCHING WAS IN MY BLOOD, although I was raised in a city for much of my childhood and adolescence. I was born in Los Angeles in 1923. From the age of six to eight, I went to a military boarding school in Northern California, in Palo Alto. During my grammar school years, I lived in the Marina District in San Francisco. I spent every free moment I could playing on the Marina Green and catching fish, crabs, and smelt in the yacht harbor. I hated being indoors, especially after my parents sent me, when I was nine, to spend the summer on a cousin's cattle ranch in Colorado.

It was in Steamboat Springs, a one-horse town surrounded by breathtaking mountain vistas. I learned to rope, ride, and brand cattle that summer, and I couldn't get enough of it. When I told my parents that I didn't want to come home, they decided to let me spend the school year out in Colorado. For many months, I had no formal education, but I learned how to punch cattle and started dreaming of owning a ranch of my own one day.

Hunting, too, ignited my imagination. That same year, when I was nine, I read *African Game Trails* by Teddy Roosevelt, his memoir about a safari he took in 1910, after he left the White House.[1] The book was filled with stories of stalking the Big Five—elephant, lion, rhino, leopard, and Cape buffalo. It was heady stuff. As I got older, I grew to love the adventure of hunting pheasants in the rice fields around Sacramento and hiking in the Sierras, living off fish I caught and a little ration.

My life, however, led to different adventures. In 1940, when I was seventeen, my mother moved to Honolulu, Her second husband, George D. Center, was a legendary swimming coach and a founding member of the Outrigger

---

1. *African Game Trails: An Account of the African Wanderings of an American Hunter-Naturalist* (New York: Charles Scribner's Sons, 1910).

Canoe Club. I soon joined them and started my junior year of high school at Punahou School in Honolulu. I had been a decent student, though not much of an athlete, in San Francisco, and I bloomed in Hawaii. I did well in my classes, had a busy social life, and set the state high school record for the bamboo pole vault.

The attack on Pearl Harbor on December 7, 1941, ended that Hawaiian idyll. As the Japanese invasion engulfed the Pacific, civilian travel from Hawaii was restricted, so in early 1942 I shipped out as an ordinary seaman in a freighter convoy to California. That summer, I worked as a carpenter on an airfield in Marysville and applied for admission, on scholarship, to Stanford as well as the University of California, Berkeley. Stanford granted me a track scholarship and UC didn't, so I enrolled in Stanford in September 1942. That November I also enlisted in the Marine Corps with a group of Stanford jocks. The Corps, in its wisdom, sent all the Stanford officer training enlistees to UC Berkeley. As a result, Cal had seven Stanford men on its varsity football team in 1943. After the war, all seven of us returned to Stanford to play for its varsity football team.

Team captains before kickoff of Stanford vs. University of California game, 1948 (author on left).

IN 1945, AS THE WAR WAS ENDING, I was commissioned as a second lieu-tenant in the Marines, and I entered Stanford Law School the following year. I maintained a scholarship and competed in track and football by not taking my bachelor's degree until my third year of law school. While I was at Stanford, I served as captain of the track team in 1946, as student body president in 1947, and as football game captain in 1948. Best of all, I met Illie—Mary Ilma Costi-gan—who had a thousand friends. Everyone loved Illie. I treasured her.

Illie and I married in 1948, and our daughter, Christen, was born the next year. After law school, I had remained active in the Marine Corps Reserve, and on June 15, 1950, I was among the first reservists called up to fight in the Ko-rean War. Exactly three months later—on September 15, when I was in Kobe, Japan—I found myself attached to the First Marine Regiment. We were sched-uled to land, that same day, in the first wave on Blue Beach at Inchon, Korea.

We had first heard about our scheduled landing a few days earlier from Kobe bar girls as we were loading our ships. "Oh, Marine, you land Inchon," they said. We were amused. We thought they couldn't possibly know where we were heading and speculated that it was disinformation. After we put to sea and opened our combat orders, however, we discovered that we were, in fact, land-ing where everyone in Kobe had said we would—at Inchon.

We were wondering what kind of reception we would receive as we prepared to ride thirty-foot high tides through harrowingly narrow straits to storm Blue Beach. Luckily, the North Koreans didn't believe their own intelligence reports; they couldn't imagine that Americans would be foolish enough to venture into Inchon's steep-walled harbor, through easily mined channels within rifle range of opposite shores. As a result, we faced only modest resistance as we landed on Blue Beach in a column, strung out in a long line through a narrow breach in a rock seawall.

The daring attack, planned by General of the Army Douglas MacArthur—against the advice of many in the U.S. military—proved to be an early and decisive victory. It was, that is, until we pushed north after retaking the capital, Seoul, at the end of the year and met 200,000 Chinese soldiers at the Chosin Reservoir, bordering China. The conflict would simmer on for years. Thanks to my early participation in the war, however, I was among the first troops sent home from Korea. In early 1951, I came back in good shape to Illie and Christie in California.

I believed that there was no place better than Hawaii to raise a family, so that year we packed up and moved to Honolulu. I remained in the Marine Corps Reserve—retiring as a colonel thirty-three years later—and continued

Colonel Martin Anderson, 1980.

my law practice in Hawaii. During our first six years there, life was wonderful. I was a trial lawyer in the firm of Anderson, Wrenn & Jenks. I also discovered my entrepreneurial streak. In the early 1950s, on the side, I became a partner in a small auto repair business. Then, with a small investment, I started a trade publishing company.

I also, by chance, got involved in the ski business in California. Illie's childhood friend Eleanor Fulstone had married a tax attorney named Hugh Killebrew, who owned about 20 percent of a Lake Tahoe ski area called Heavenly Valley. Hugh told me that his partners—a couple of major U.S. corporations—were trying to squeeze him out just as the ski industry was beginning to boom. I wanted to help Hugh, and Heavenly Valley looked like a good opportunity. So I contacted my friend Rudy Peterson, the former president of Bank of Hawaii, who was by then president of Bank of America. With a generous loan that Rudy arranged, I bought out Hugh's partners, gave Hugh a majority share, and kept a large stake for myself. I was learning that I loved action, not just legal transactions. It was a charmed life.

Then, one morning in 1958, Illie's doctor discovered a tumor in one of her legs. After performing surgery to remove it, he told me that he thought he'd have to amputate the leg. He wouldn't know anything for sure until the biopsy came back from the mainland. Still, he wanted to prepare me. He didn't mention graver possibilities, and I couldn't imagine them.

My face must have looked ashen when I left the hospital and returned to the office. As soon as senior partner Heaton Wrenn saw me, he said, "What's wrong, Marty?" I gave him the news.

Three days later, when Illie's results came back, we learned that her tumor was benign. She could keep her leg, but there was considerable damage to the muscle. She would need to have physical therapy and vigilantly exercise the leg for the rest of her life.

With fortune smiling on us for the moment, I resolved to take a break from my long workdays. I told Heaton that when Illie recovered completely, I wanted to take off on an extended two-month vacation. "There is a lesson here, and I won't forget it," I told him. "I want to take Illie to all the great art museums that she's never seen—the National Gallery, the Louvre, and the Prado. I want to show her the world's most beautiful paintings and take her on the trip that she's always dreamed of."

"Take whatever time you need," Heaton told me. "In fact, take two months to travel around Europe and show her the places she's always wanted to see. Then," he added, "I want you to take a third month and go on safari in Kenya. I'll tell you exactly how to go about it."

It was an adventure that would change our lives.

CHAPTER TWO

SAFARI

THE PLANNING FOR OUR TRIP took an entire year, and Heaton dispensed much advice, over numerous lunches, about how we should equip and organize our safari. I still remembered the details of Roosevelt's 1910 expedition—a lavish affair with some 250 porters and guides. The former president had the backing of Andrew Carnegie, the scientific sponsorship of the Smithsonian Institution, and the financial assistance of the publishing house Charles Scribner's Sons, which would publish the memoir of his eleven-month journey across Africa.

Our safari, needless to say, would be a lot more modest. I was doing well in the law, but cost was a consideration. Besides, we were bringing along nine-year-old Christie. We wanted a family vacation and would certainly not be "dressing for dinner" in the bush or expecting to encounter the liquor and luxury common on safaris since the 1930s. We needed to find an outfit that could meet our particular needs, for a reasonable fee. And that was a challenge.

I first noticed the name Michael Prettejohn in a copy of *Outdoor Life* magazine, in a small advertisement for his safari services. Soon after, I read an article about the twenty-six-year-old Kenyan in the June 1959 issue of *Look* magazine, which detailed the political changes sweeping his country, which had been a British colony, governed by the British Crown, since 1895. The capital of Kenya, Nairobi, had been a watering stop for "the lunatic express," the rail line the British government had built from the Port of Mombasa to Lake Victoria. Many English settlers came to Kenya after World War I—mainly veterans who had been recruited to establish small farms in the fertile region, known as the White Highlands, around the Rift Valley and from Nairobi north to the frontier. The region's high elevation assured a healthy environment, and the British government encouraged the migration by promising that the White Highlands "would be forever England."

Over the decades, as the white settlers grew in number—from about 100 in 1902 to 80,000 in the early 1950s—they increasingly clashed with colonial officials over export and agricultural policies.

After World War II, winds of change swept Africa as colonial rule was coming to an end. In 1952, members of Kenya's Kikuyu tribe led a violent backlash against the government. The uprising, called the Mau Mau rebellion, caused the British governor of Kenya to declare a state of emergency, which was followed by arrests and reprisals against suspected rebels. In 1956 the uprising was quelled, but it was clear that Kenya was moving toward independence. In 1959, the colonial government stunned European settlers by proposing to open up the exclusive White Highlands to African and Asian landowners. In a country where blacks outnumbered whites ten to one, African Kenyans—led by twenty-eight-year-old Tom Mboya from the Luo tribe and its chief, Oginga Odinga—were calling for "one man, one vote" government.

Look described Mike Prettejohn—a third-generation settler—as the young white Kenyan counterpart of Tom Mboya. Born in Nakuru, northwest of Nairobi, and raised in his early childhood in a mud-and-wattle hut on the edge of the Rift Valley, Mike was called up by the British military offensive against the Mau Mau. With ranchers and farmers, and aided by many native Africans, he had helped battle 20,000 insurgents in the forests and jungles of the rugged Aberdare Mountains. "My future is here," Mike declared in the article. His appealing determination and his love of Kenya persuaded me to hire him as a guide.

Mike had grown up when the country was untamed and teeming with wildlife. It was pure paradise for a boy with a taste for hunting. At age six, he got his first air gun and shot pigeons for the pot. At eight, he was using a .22-caliber rifle to hunt rabbits and steinbuck. He shot his first buffalo at age sixteen, and at eighteen he was hunting elephant, rhino, and other big game on license or doing control work for the local game warden.

Mike had spent his early years with his parents on the 30,000-acre ranch owned by his step-grandfather. His family then moved to the 60,000-acre cattle and sheep ranch of Lady Eleanor Cole, an early British advocate for African and women's rights in Kenya. After graduating from the Prince of Wales School in Nairobi, Mike wanted to be a hunter or a ranger with Kenya's Game Department. His father, Joe Prettejohn, however, wanted him to buy a farm property and settle down. At the time, Joe Prettejohn was chairman of the European Agriculture Settlement Trust, which provided financial aid to settler farmers in Kenya after World War II. The trust loaned Mike enough capital to buy Laburra, a 3,000-acre property between Mount Kenya and the Aberdares. Laburra had small flocks of sheep, goats, and cattle; abundant wildlife; easy access to forests filled with big game; and a magnificent old cedar house with an outstanding view. Laburra's herds and few acres of wheat were tended by a dozen families from the Kikuyu tribe, whose native huts were near the sprawling main house and attached guesthouse.

Mike planned to repay the trust by basing hunting safaris at Laburra. Another, more profitable opportunity, however, soon emerged. Shortly after Mike bought the property, film scouts visited the ranch, declared it ideal for their projects, and leased Mike's house for two movies that netted him enough money to repay the entire loan. One of the movies was about the Happy Valley set, an infamous crowd of promiscuous British aristocrats in Kenya in the 1920s and 1930s. The other film was *West with the Night*, based on aviator Beryl Markham's memoir about her life growing up in British East Africa at the turn of the century. A year before I contacted Mike, he had become a licensed professional hunter, and as it turned out, we would be among his very first clients.

I had been rated an expert rifleman in the Marines, and as our safari plans began taking shape, I became more and more excited about the prospect of hunting the Big Five. I couldn't wait to get to Africa; it was practically all I could think about while we traveled through Greece, Italy, Spain, Germany, England, and Egypt in the summer of 1960.

AT LAST, IN THE EARLY AUTUMN, we arrived in Kenya. Illie and I were transported by its beauty and the astounding wildlife. In the mornings we would walk trails in the nature reserves near Laburra or go north into sparsely inhabited bush to hunt leopard and buffalo. We'd spend our afternoons listening to Mike's stories of the country's nomadic tribes—the Samburu, the Suk (or Pokot), and the Rendille—and tales about the British settlement of Kenya. In the evenings, we'd enjoy the warmth and companionship of hospitable neighbors.

It was obvious, though, that while Illie, Christie, and I were captivated by this exotic, wild, and magnificent place, the English settlers we met were in a state of anxiety. Many were beginning to think about where they might go if African Kenyans won control of the government.

"Have you sold yet?" was a common greeting that colonialists exchanged when they saw each other. They were shocked by what they viewed as England's

Christie, Illie, and Marty before leaving for first trip to Kenya. 1960.

Illie, Marty, and Christie with Mount Kilimanjaro at the beginning of our Kenya adventure, 1960.

Cedar ranch house at Laburra/Sangare, 1960.

betrayal; they could hardly bring themselves to believe that Her Majesty's government would open up the White Highlands. Some believed it wasn't too late to stop change in Kenya or that the ultimate reckoning was a decade off. Still, every discussion centered on the future and the problem of what they should do. Some adopted a wait-and-see attitude. Others insisted they wouldn't sell out and leave, but many—especially those who couldn't tolerate the idea of being ruled by Africans—advocated selling out and moving to Rhodesia, South Africa, or back to England.

At the time, Mike's father was organizing trips for British settlers to view South African properties. Many were signing up, including those who were attracted by the clarity of apartheid. H.P. also announced, in his role as chairman of the Settlement Trust, that he would make sure that all British farmers in the district who gave up their land would be paid out in the United Kingdom. He had already begun to identify parcels of property on which African Kenyans could be settled, and he offered to arrange for Mike to leave with compensation.

Over and over again, Mike talked with us about whether he and his family should abandon Kenya for South Africa, as his parents and brothers planned to do. My enthusiasm for Kenya, I admit, led me to encourage the Prettejohns to stay in their country instead of leaving. Once Mike introduced us to a high-spirited young rancher named Gilfrid Powys, I felt even more positive about Kenya's future and hopeful that Mike and his family would decide to stay.

Gilfrid, a tall, athletic twenty-two-year-old, gave us a tour around one of his family's farms and talked exuberantly about Kenya and what it offered. His father, William, had come to British East Africa from Somerset, England, in 1913 to forge a life as an early pioneer rancher. After he served in the East African Mounted Rifles during World War I, the British Government had given him several thousand acres, at Kisima in the Rift Valley.

Gilfrid's mother, Elizabeth, had served during the war as a British VAD (Voluntary Aid Detachment) nurse in France. The British government had also allocated her a soldier-settler farm for her distinguished service. Elizabeth was the Powys family's thinker and politician. She was also an advocate of Kenyan independence, an ardent hunter, and an active member of the Capricorn Africa Society, which worked to end discrimination in central and East Africa.

Gilfrid had been born in a mud hut on the slopes of Mount Kenya. After boarding school in Nairobi, he had studied at Gordonstoun in northern Scotland. After graduation, he returned to Kenya, where the colonial government recruited him to help African families in the native reserves after the Mau Mau rebellion. It was a rewarding job for a year. Gilfrid then managed one of his family's farms and joined up as a mercenary with the Sultan's Armed Forces of Oman, tracking terrorists who used camels to smuggle land mines into the country. A few months before we met, Gilfrid had returned to Kenya. He was, he said, committed to his country's future, and he had no patience for racist settlers who dreaded the prospect of Kenyan independence.

With our shared love of hunting and our military backgrounds, Gilfrid and I spoke the same language. At the end of our visit, he turned to me and said something that would change my life: "The country needs a few of your sort, Marty. I'll sell you a piece of my family's farmland if you decide to join us."

Gilfrid's words planted in my mind the possibility of owning a ranch in Kenya. At the time, I was actively considering purchasing a ranch in British Columbia—if I could find a local partner who could run it while I continued my law practice. Now the idea of spending part of my life in Kenya was an intriguing prospect.

A few nights later, Mike and I sat around a campfire, again discussing the decisions that he had to make. If he abandoned Kenya for Rhodesia or South Africa, I suggested, he would only be postponing political problems that his children, Giles and Jessica, would have to face. I knew that Mike was leasing

2,000 acres on Laburra's western boundary, and I suddenly suggested that I buy the property. "We could go into partnership, Mike. We'll do it fifty-fifty."

"That's easy for you to say," he replied. "You'll put in some money and go off and leave me with all the problems."

"No," I insisted. "I believe in this country, and I want to spend more time here."

Mike wasn't convinced, but he agreed to think about my proposal.

CHAPTER THREE

THE ELUSIVE BONGO

SOON AFTER, MY FAMILY RETURNED TO HAWAII, and Mike visited South
Africa to see what opportunities he might have there. Within a couple of months
he made a decision.

"My place is in Kenya," he wrote to us. "I am not going to sell Laburra. I am
going to give it a go with you and stay on." Mike said he wanted to work with his
countrymen to forge a new, racially integrated nation, and he was giving serious
thought to my idea of a partnership. Mike's wife, Gill, however, was understandably cautious about my proposal. "After all," she asked Mike, "what do you really
know about Marty? What does this American lawyer really have in mind?" In
the end, though, Gill gave her blessings to the plan and became a close friend.

In 1961 Illie and I bought the 1,500-acre farm next to Laburra and created a
partnership with Mike and Gill. We had now committed a part of our lives to
Kenya. We were landowners with a concrete reason to return there and spend
time with the Prettejohns. Over the next year, we strengthened our bond with
them through an energetic correspondence. Mike and I worked out the business details, and our wives kept the families abreast of everything.

Hunting wildlife, however, was still my main interest, outside of my law practice, and in August 1962 Illie, Christie, and I returned to Kenya for another safari.
This time, with Mike's help, we divided our thirty-five-day visit into three parts.[1]
First we went south into elephant and rhino country on the border of Tsavo
National Park. Then we went into the Masai, near Narok, where we hunted lion,
buffalo, Grant's gazelle, and other game. Finally, we spent the last week and a
half of our trip at Laburra. Illie had lots of opportunities to fill her sketchbooks
with drawings of wildlife, flowers, and camp life; twelve-year-old Christie and a

---

1. Martin Anderson, "Bongo on a Family Safari," *Karatasi Yenye Habari*, quarterly of
the Shikar-Safari Club, published in Cary, Illinois (Summer 1963): 6.

friend from Kenya had contact with the colorful tribal life of the Masai people; and with the Prettejohns' farm as a base, Mike and I were able to hunt bongo—a rare nocturnal antelope—in the dense bamboo forests of the Aberdares.

Bongo was a coveted trophy. The large animal had spiraled, lyre-shaped horns and a chestnut hide beautifully patterned with white stripes. When it came to finding bongo, however, the odds were against us. Only five had been shot on forty-one licenses in 1961 and just two on thirty licenses in 1960. Mike and his skilled African trackers, however, had found signs of bongo activity in several locations in the Abedares, so we set off into the mountain range.

The weather was unseasonably cool and wet, and the ground was soaked. Mike grinned as we made our way up a slippery red clay game trail. "This rain is a blessing," he said to me, "because the bongo should be dropping down out of the high forest bamboo to dry off a bit."

Walking beneath a canopy of tall, thin trees, we came across deep buffalo and rhino tracks. Soon the head tracker stopped, stooped, parted the grasses, and pointed at another fresh set.

"Mguu Dongoro," he said in Swahili; it was a bongo. We had no intention to track the animal, however. Our plan was to find an area frequented by a bongo bull, then return and camp overnight at the site later in the week. We would rise at first light and catch the bull as it moved back into heavy growth after the night's feeding.

Two days later, we set off from Laburra again—this time to spend the night where we'd spotted bongo tracks. That evening, when we stopped after our long trek, the dark dropped quickly, and the forest came alive. Cape buffalo grunted near our campsite, and three giant forest hogs slogged down the game trail, halting only a few feet from our bedrolls. A bushbuck barked a warning of a stalking leopard. The rustling of game and calls of the night birds pierced the mist that masked the African stars.

When we arose just before dawn, the forest was silent, as if the night creatures had never been there. A light morning breeze stirred the treetops, and just enough light filtered through the mist and leaves for us to find the game trail. We followed it up a narrow valley, walled in on both sides by towering vegetation. Suddenly the lead tracker froze and stared down at the fresh tracks of a big bull. The wind was right, and heavy dew muffled our footsteps as we followed him. A few minutes later we slowed our pace as we approached a clearing. Parting the bush, I saw a scene that looked like it was out of a dream. In the early-morning mist, surrounded by walls of forest green, I saw a set of magnificent lyre-shaped horns, carried by a huge, majestically striped animal.

Marty with rare bongo bull from the Aberdares.

"It's your bongo!" Mike whispered to me.

Mechanically, I drew a .30-06 rifle to my shoulder. When I saw the cross-hairs on the animal's thick neck, seventy-five yards away, I fired and sent the Nosler bullet home. The bongo fell, and the silence exploded with our back-pounding excitement. I'd caught the elusive bongo. The giant's graceful horns measured thirty-four inches in length.

My friendship with Mike, under the trying conditions of this safari, was completely cemented. In that rough country, we got to know each other's personality and guts. We never discussed it, but we knew that we could rely on each other. It was a partnership that would direct the course of our lives for nearly three decades.

CHAPTER FOUR

WHITE SALE

ONE YEAR LATER, Britain ended its colonial rule of Kenya. The country had negotiated an independent Constitution—though it remained a member of the British Commonwealth—and in May 1963 Kenyans voted in their first general election. The Kenya African National Union (KANU), which advocated a centralized government, won 83 of 124 seats, defeating the Kenya African Democratic Union (KADU), which favored an ethnically divided federal state. On June 1, KANU's leader—seventy-three-year-old Jomo Kenyatta—became the country's first prime minister. After the adoption of the country's constitution, he was designated president.

Kenyatta had lived and studied in England for more than fifteen years. After the Mau Mau rebellion in 1952, he had been arrested with five other KAU leaders. They were all charged and tried for allegedly leading the Mau Mau movement. Although Kenyatta was a moderate and had publicly denounced Mau Mau, he spent the next nine years in prison and detention. Finally, after more than a million Kenyans demanded his release, Kenyatta was freed in August 1961. Now gaunt, white-bearded, and dignified, he led the new independent nation of Kenya.[1]

Kenyatta's message was one of unity and reconciliation. "Let there be forgiveness," he declared. His rallying cry was *"Harambee"*—Let's work together—and he urged white settlers to stay in Kenya for the good of the country. He even addressed anxious colonialists on their own turf, in a town hall meeting in the White Highlands. Kenyatta assured them that the new government would not take their property. "Let us join hands," he said. "We want you to stay and farm well in this country. That is the policy of the government." Many in the

1. For a comprehensive treatment of the Mau Mau insurrection, see Part II of Ian Parker, *The Last Colonial Regiment* (Librario Publishing, 2009).

audience who had previously been hostile to Kenyatta responded with shouts of *"Harambee!"*

Not all settlers, however, trusted or embraced Kenyatta. A fourth of them—20,000 Europeans—put their farms up for sale, took their savings out of the banks, and packed their bags. Mike Prettejohn's neighbors, the Sutcliffes, were among the many who planned to flee. They were selling their 3,000 acres of rough, beautiful land—including a 20-acre property and lake adjoining Laburra—for £3,000. It was an opportunity to expand our holdings that might not come again. I suggested to Mike that I pay up front for the property; he could later pay me back for his half, and we'd be equal partners. Mike and I purchased the Sutcliffe parcel, consolidated it with our other land, and called the entire property Sangare.

As settlers left Kenya, we also had great opportunities to buy stock. When a departing rancher near Nairobi offered a mob of steers for a quick sale, Mike bought them at a bargain, then sold them within a year for enough profit to pay me back for his share of the Sutcliffe lands. More large properties, too, were becoming available, though some of the best parcels were too pricey for my bank account. When a ranch with 60,000 acres and 6,000 head of cattle came on the market for £100,000, we had to pass on it. But soon a 40,000-acre property, the Luoniek Ranch, came up for sale. Located in the remote northwest corner of the Highlands, the property had lion on it, Samburu neighbors to the north, and Suk neighbors to the west. Its English owner was so anxious to leave that all he wanted in exchange for the property was the cost of his livestock and animals. It was an attractive offer. Still, before we plunged ahead with the deal, I wanted additional assurances from the Kenyan government.

I wrote to J. H. Butter, the minister of finance in Kenyatta's shadow cabinet. Butter assured me that approved investments made in dollars and other hard currencies were safe. The government was encouraging overseas investment, because the country's development depended on its ability to attract outside capital. If we paid for the property in dollars, he told me, the government of Kenya would give the land purchase approved status as a foreign investment. The capital I invested in the purchase, moreover, would then be insured by the government of the United States, which was doing everything it could to encourage Americans to invest in Kenya. With those assurances, I purchased Luoniek and its 2,000 head of cattle.

We had some concerns, however, about being able to keep all of Sangare, given the government's plan to resettle landless African families on selected parcels in the White Highlands. As a preemptive move, we decided to sell the first property we had purchased, adjoining Laburra, to Kenyatta's forty-two-year-old son, Peter Muigai Kenyatta. It was a manageable farm, and selling the parcel was a sign of good faith and cooperation with the new government.

Meanwhile, we hired one of Mike's old schoolmates, Humphry Clark, to manage the distant Luoniek ranch. Humphry was an ideal choice for such a remote place. A lean loner who had never been married, he had capital to invest as a partner, since he had recently sold his family farm at the edge of the Abedares.

In November 1964 Humphry sent me upbeat news of our progress at Luoniek, reporting that we had purchased two hundred head of young stock, for ten pounds a head, with help from Gilfrid Powys. On a troubling note, however, he wrote that our Suk neighbors were grazing their cattle on our land, depleting grass we needed for our own herds.

A month later, I wrote to Bruce Mackenzie, Kenya's minister of agriculture. I wanted to advise him, as a foreign investor, of our substantial progress on the Luoniek ranch. Mackenzie—an efficient, indefatigable Scottish South African who didn't suffer fools gladly—was the only white member of Kenyatta's cabinet. His connections, informality, and breadth of knowledge were impressive, but he was truly formidable because he had the president's ear. I informed him that trouble was brewing over water rights on our boundary with the Suk. Mackenzie, I knew, was already aware of problems with the Suk herders in the area, since he had been trying to resolve a grazing dispute between the Suk and another property.

Over the next six months, Suk trespassing worsened because of a severe drought. When Humphry and Mike flew over Luoniek, they spotted herds of Suk cattle grazing on our property. "If not stopped very soon," Humphry warned, "[the Suk cattle] will have eaten all our grass in that area [and] leave little for us." We also experienced a deadly confrontation with Samburu stock thieves. A group of Samburu raiders beat one of our herders so badly that Humphry had to rush him to the hospital. Four hundred of our steers also went missing, although we later recovered them.

To protect the Luoniek ranch, I suggested that we propose a regional security patrol. If our neighbors didn't want to participate, I was in favor of hiring a few men on our own to keep the raiders and trespassers in check. The situation, however, was worse than I thought. A local representative of the government's agricultural committee visited Mike and confessed that the government could not restrain the Suk and protect our land. Instead, he suggested, the government might be willing to buy us out of Luoniek and move our operation to some holding land until another suitable property could be found to replace it. Mike telephoned me in Hawaii—a rare occurrence in those days—to report the official's visit, and the news shocked me. The outlook had to be hopeless for the government to make that admission. I wrote to Bruce Mackenzie again, to express my surprise and to stress that Mike and I were anxious and determined to make a contribution to Kenya by running a successful cattle operation. Unfortunately, I told him, we no longer knew where we could do so.

Soon afterward, the local agricultural representative came to see Mike again, this time with a proposal. The government, he said, was planning to offer for lease an enormous tract of untamed, mostly uninhabited, semi-arid land north of the Galana River. To develop Kenya's cash-starved economy, Kenyatta was eager to develop vacant, underutilized land where no African Kenyans wanted to settle. Other groups were already expressing interest in the Galana tract, but the representative assured Mike that if we applied for the concession, he and Mackenzie would give us prominent consideration.

On my next visit to Kenya, Mike and I went to see Mackenzie in Nairobi. He outlined the basics of the deal the government wanted for the Galana lease. The project would require an investment of $100,000 and the development, over time, of a ranch carrying 26,000 head of cattle. He acknowledged—given all the tsetse fly, East Coast fever, and other tick-borne diseases in the area—that there was no guarantee cattle could survive. The Galana lessee would also have to develop water holes, roads, and a causeway over the Galana River on land where there was now virtually nothing and no one, except for a small tribe of people known as the Waliangulu. These were gentle people, Mackenzie stressed, "except that they kill elephant with poison arrows."

Government rangers, Mackenzie added, were also shooting three to four hundred elephant a year in the Galana area. Officials believed that they had to crop the herds that moved back and forth between Galana and the adjacent Tsavo National Park, because the growing elephant population was crowding the park and destroying its tree cover, their source of browse. The government, in the future, wanted to send the meat of the cropped elephants to Nairobi to help feed the *wananchi*—the Kiswahili word for citizens—Mackenzie said.

If we wanted to move forward, he explained, we should submit an offer of tender. After our offer was reviewed, we would face a series of interviews by different ministries that had an interest in developing the area. We would likely have competition, he added, from a Texas oil millionaire and an Italian filmmaker who wanted to use Galana for film locations. There was also a group of five English settlers who thought they had the inside track, "but I don't like them, and they don't like me," Mackenzie told us.

We left the meeting intrigued. Mike cautioned me, however, that the property and project were too large and ambitious for the two of us to handle alone. He suggested that we ask Gilfrid Powys to join us in the venture. Gilfrid, he told me, had recently spent six months at Galana filling in for his childhood friend Ian Parker, who had been serving as the Game Department official in charge of the area since 1960. I liked the idea of bringing Gilfrid in on our plan, so we flew up to his ranch to try to recruit him. It was sheep-shearing season, and we found him sitting on a bale of wool, taking a lunch break in the shearing shed. Mike and I shared his tin of bully beef and flask of coffee, and we showed him the advertisement inviting tenders to develop Galana. We asked him to join us in applying for the lease, and he said, "Yes, why not? But first, you're going to have to go see the old man."

So Mike and I went up to see Gilfrid's father, who was one of the biggest, most respected European landowners in Kenya. We found Will Powys, who was then in his eighties, out shearing sheep. We told him about the Galana project, and his eyes lit up.

"I'd love to have Gilfrid break a bit of Africa like I did when I came out here at the turn of the century," he said. "We'll do it."

The stars were aligning nicely. Gilfrid knew all the key officials in the Galana area, and he knew the land. Mike, Gilfrid, and I—a hunter, a rancher, and an American lawyer—could make a great team.

That afternoon, we confirmed the government's purchase of Luoniek. Cheered by the news, we bought a new Cessna 180 airplane to replace our old Tri-Pacer. Three weeks later, we flew down to Galana and made an aerial tour of the 1.6- million-acre tract—surveying the water holes, grazing areas for cattle, and the bush that we would keep for wildlife. After we landed, we hired an old Land Rover from an Arab cattle trader to explore the property. Galana's single roadway had been built forty years earlier by Denys Finch Hatton.

Everywhere we went over the next couple of days, we saw rhino, elephant, lesser kudu, eland, oryx, and other game. The property was extraordinary and vast. My excitement about Galana, however, was tempered by our Waliangulu guide, Wario.

"Why do you want to come here?" asked Wario, a tall, serious, distinctive-looking man. Galana had never been of much interest to anyone but the Waliangulu, and Wario seemed worried that we might ruin it. We tried to explain to him that sooner or later someone was going to develop the land, and that it would be better for all concerned if we were chosen.

After Wario left us, his question got to me. Was developing this virgin land the right thing to do? Would we destroy it for the elephant and the Waliangulu? If we didn't take the lease on Galana, who would?

Ultimately, the prospect of being stewards of such uncommon land eclipsed our concerns. We were enthusiastic about the possibilities of running an enormous cattle ranch, experimenting with domesticating wildlife to help feed Kenya's people, and controlling a hunting sanctuary where we could protect the elephant from excessive cropping. Mike, Gilfrid, and I wrote a three-page application for a forty-five-year lease of Galana and submitted it to Bruce Mackenzie. Then, without much faith that the deal would work out in our favor, we went back to our lives—Mike to his safaris, Gilfrid to his sheep, and I to my legal practice in Hawaii.

CHAPTER FIVE

POISON ARROWS

SEVERAL GROUPS OF BIDDERS LINED UP as the government weighed the applications to lease Galana. Our leading competitors included a settler group organized by Cedric "Titus" Oates and Ian Robson. They had been involved in a large operation known as Laikipia Ranching Company and had the backing of an investor in the United Kingdom with cattle-ranching interests in Argentina. Oates had come to Kenya years earlier as a government agricultural officer and knew Galana well from the days when he was posted to the region. Like Bruce Mackenzie, he had a strong—some might call it quarrelsome—personality. It was little surprise that he and Mackenzie were at odds.

The other competitive bidder was Ray Ryan, a storied American oilman, a gambler, the developer of El Mirador Hotel in Palm Springs, and the owner of substantial properties in Southern California. In 1959 Ryan had partnered with Oscar-winning movie star William Holden and Swiss banker Carl Hirschmann to renovate the old Mawingo Hotel at the foot of Mount Kenya and turn it into a club for the rich and famous. In association with 20th Century-Fox, Holden had built a $100,000 soundstage in a secluded quarter of the club's 90-acre grounds and made the movie *The Lion* there. Ryan had purchased two adjoining ranches in the White Highlands, totaling 52,900 acres, and was, according to *Sports Illustrated*, in the process of stocking the land with 10,000 head of cattle. He had also bought Zimmerman Ltd., the famous Nairobi taxidermy firm, and created the Ryan Investment Company to back other projects in Kenya.

Ryan and Holden, according to some accounts, were hoping to become two of the biggest *bwanas* in the country. They gave charter membership in their club to an impressive list of figures, including Jomo Kenyatta, Tom Mboya, former U.S. president Lyndon B. Johnson, Prince Bernhard of the Netherlands, U.S. senator Everett M. Dirksen, and American celebrities including Bing Crosby, Bob Hope, Walt Disney, and John Wayne. By early 1966, however, Ryan dropped out of the bidding for Galana—most likely because he was unhappy with the government's insistence on controlling all of its hunting quotas and revenues. The management of wildlife at Galana, in fact, was a politically complex issue. Naive and eager, we had little understanding of competing interests and agendas and how powerfully the economics of land use would come into play.

The Galana area had first attracted government attention in 1956, at the end of the Mau Mau uprising. As white Kenyans were beginning to refocus their attention on wildlife issues, two game wardens—David Sheldrick and Bill Woodley—exposed the fact that there was a devastating amount of illegal poaching in Africa's largest game preserve, Tsavo National Park. The colonial government responded by assigning Sheldrick and Woodley to tackle the poaching problem in the eastern portion of Tsavo.

The most prolific poachers were the Waliangulu, known officially as the Waata. The tribe was so small that it had been virtually unknown as a distinct society before the revelations about poaching. Its history, however, was impressive. The Waliangulu—meaning "tortoise eaters" or "snail eaters" in Bantu slang—were thought to be the original inhabitants of the Tana River. They were possibly direct descendants of ancient hunting people depicted in Neolithic cave drawings, and they had hunted game—elephant, in particular—from time immemorial.

The excellent Waliangulu hunters used distinctive longbows with a draw power of more than a hundred pounds, greater than the draw power of medieval English longbows. All one needed to do was to look at the musculature of the backs and chests of the men who hunted with these weapons to understand how much force was required to flex and fire the bow and how much practice those actions must have required.

The Waliangulu armed their bows with heavy, poison-dipped, lethal arrows. They manufactured the toxin by boiling the bark and leaves of certain types of *Acokanthera* trees for about seven hours. The process produced a tarlike substance containing an extremely poisonous glycoside, known as oubain, which the tribesmen smeared on the arrow's shaft and head. Because the toxin's potency diminished with exposure over time, they applied it to the arrows shortly before a hunt. The Waliangulu tested the potency of the poison, known as *Hada*, by smearing a small amount on a thorn and pricking a frog's flesh with it. If the poison was effective, they stored it in plant husks until they used it.

The hunters wasted little effort in tracking elephant. They would lie in wait by a watering hole or a well-used elephant path, hoping to be within twenty

Waliangulu hunter.

feet of the tusker when they fired their arrows. If possible, they aimed for the elephant's spleen, so the poison would be absorbed into its bloodstream as quickly as possible. If the arrowhead lodged deeply and accurately, the toxin would go to work almost immediately and could kill a 12,000-pound elephant within minutes. If the arrow didn't perfectly hit its mark, the animal could take days to die.

In the 1950s, poaching by the Waliangulu inflicted immense losses on the Dabassa elephant herd, which moved between Tsavo and Galana. David Sheldrick, in one search between the Galana and Tana Rivers, discovered the carcasses of about 900 elephants. He also recovered

352 tusks, weighing more than 6,500 pounds. If not for Sheldrick and Woodley's two-year anti-poaching campaign, the tribe might have destroyed the area's elephant population. The two rangers organized a paramilitary operation, employing a field force of fifty men, that carefully cultivated and depended on an informer network. When Sheldrick and his men caught a poacher, they offered the prisoner freedom or a reduced sentence in exchange for cooperation.

Once the anti-poaching forces had acquired enough information, they staked out a community that harbored active poachers and raided it in the middle of the night. Dragging the poachers from sleep, they confronted them with an arsenal of incriminating evidence discovered during the raid and detailed information that they had gathered from interrogations. Drowsy and disoriented, the poachers frequently implicated themselves and others they believed had betrayed them.

At the beginning of the campaign, the poachers would provide extensive information when they were threatened with long sentences. Later, however, they learned to hold their tongues, keep evidence of poaching away from their homes, and plead ignorance if they were arrested. The cat-and-mouse game between the wardens and the poachers became more challenging, but the campaign was successful enough that by 1957 Sheldrick was able to anticipate a time when elephant numbers would increase again in Tsavo East.

The anti-poaching effort saved the elephant population in the 16,000-square-mile ecosystem, including the Galana area, but it created other complications. Most importantly, the Waliangulu were, in effect, now unemployed and unable to make use of their traditional skill. They needed a way to earn a livelihood, but only a limited number could be put to work as gun bearers and trackers for professional hunters, and even that work was inconsistent.

This issue was of great concern to Sheldrick, Ian Parker in the Game Department, and Noel Simon, who was then the chairman of the Kenya Wildlife Society. Working together, the three men proposed that the government adopt a Galana River management scheme. Their plan would allow the Waliangulu to crop elephant, under close supervision and on a sustained yield basis, between the eastern boundary of Tsavo East and the coast. Under the scheme, the Waliangulu would be able to continue to live in their traditional way without breaking the law or depleting the elephant population.

Kenya's colonial government accepted the plan in principle, but added a requirement that would doom it. According to the approved scheme, the meat and hides of cropped wildlife could be marketed, but all revenue derived from private hunters, safaris, and the sale of ivory or rhino horns had to be deposited directly into government coffers. The operation would be deprived of those funds, but it would still have to prove that it was self-sustaining.

The scheme proved to be wildly impractical. "At the outset," Parker wrote in his book *Ivory Crisis*, "I did not think that it was going to be difficult to achieve our annual quota of two hundred elephants. But then up to that time, I had never tried to hunt wild animals to meet a budget or a schedule, or fully to use their products—tusks, meat and hides." For the Waliangulu, moreover, hunting to meet a quota was completely different than hunting at whim, on foot, and taking only as much meat as they could carry. As David Sheldrick's wife, Daphne, observed, the tribal hunters were not prepared to work consistently, on a sustained basis. After they were paid, they would periodically retire to their villages to enjoy their earnings.

Getting to the hunting grounds was also a vexing problem. According to Parker, rangers spent less time hunting than making roads through the property. The roads themselves could be very dangerous. The land in the area was so dry and featureless, especially in the north, that there was a serious danger of becoming lost on the roads, running out of gas, and dying of thirst. Visibility was poor—fifty yards or less—because the terrain was densely covered by low *Commiphora* woodlands. Parker had no aircraft to scout for game. The elephants, moreover—after years of being hunted by bowmen—had evolved intelligent means of avoiding detection; they fed downwind and approached water only at night. As Parker concluded, "It is not easy to kill large numbers of elephants on foot in thick vegetation and even more difficult to do so to a predetermined schedule."

Financially, too, the Galana management scheme was a failure. If the profits from safaris were counted, the operation was producing nearly double the amount required by the government. But because that money was credited directly to the government, the scheme was officially in the red after nearly four years, and the colonial government was preparing to shut it down.[1]

In 1961, new complications arose. Three straight months of heavy rains, followed by flooding, washed out all the roads except the original ones that had

---

1. For this chapter and the history and elephant-hunting culture of the Waliangulu, or Wata, I also drew on Ian Parker's *Ivory Crisis* (London: Hogarth Press, 1983), 24–51. Parker's book was a valuable source for the history of the original Galana River Management Scheme. For a description of the Waliangulu people and the elephant during several years before we took over the Galan area, see Dennis Holman, *The Elephant People* (London: John Murray, 1967).

been built in the 1940s. Grasses that were rarely higher than two feet were now eight feet tall, and when they dried out, someone set a torch to them. A thousand square miles of dense, spiky woodland burned to ash, leaving wide grass plains with open vistas. That changed the equation for Galana in two critical ways. It made elephant hunting easier and large-scale cattle ranching practical—if tsetse flies, ticks, and disease didn't make it impossible.

In 1963, with that possibility in mind, Parker hired Gilfrid Powys as his substitute and took a sabbatical to write a final report to the new, independent Kenyan government. He proposed that private enterprise be invited to take over the scheme and raise cattle on Galana's now-open plains. Kenyatta's government adopted his plan and subsequently advertised for tender offers. That was when we came into the picture. By 1966, once Ray Ryan dropped out of the bidding, Mike, Gilfrid, and I were going head-to-head against the settler group for the Galana lease.

Serendipitously, my work as an attorney in Hawaii gave our group an unexpected advantage. In the early 1960s, I represented a faction of the trustee owners of the *Honolulu Star-Bulletin*, led by U.S. Army Major General Edmond H. Leavey. The son-in-law of the late Wallace R. Farrington—the paper's former publisher and onetime territorial governor of Hawaii—Leavey had enormous influence. During World War II, he had been U.S. Army Forces chief of staff in the Western Pacific and had had the privilege of accepting the surrender of the Japanese military to the Allied High Commission in the Philippines in 1945. Leavey went on to become U. S. Army controller and assistant chief of staff to General Dwight D. Eisenhower during the organization of Supreme Headquarters Allied Powers in Europe. In 1952, he was elected vice president of the International Telephone and Telegraph Corporation, and he served as its president from 1956 to mid-1959. In the late 1950s, Leavey told me, he had been commissioned by Kenya's colonial government to lead an economic survey of the country. He subsequently authored a World Bank report on how to help jump-start Kenya's economy and raise the fledgling nation's standard of living.

When Mike Prettejohn and I were marshaling our forces in 1966 to secure the Galana lease, Leavey volunteered to contact former colonial officials on our behalf, among them J. H. Butter, who was still in the government. In support of our tender offer, Leavey wrote a glowing endorsement of our financial credibility. Things seemed to be falling into place: Leavey's blessings, the government's interest in foreign investment, the local participation of the Powys family, and the antipathy between Mackenzie and the Oates group.

Trouble erupted, however, once some officials realized that our tender offer was the front-runner. "At the last Cabinet meeting," Mike wrote me in early December, "every minister had his own ideas of what should become of the Galana Scheme. [One minister suggested] that it ought to run by a legal cooperative. However, I gather Bruce Mackenzie has said 'balls' to this and in exactly those words."

After this meeting, Mike added, we were contacted by the Oates group; "I think they suddenly realized that they were losing, and they've asked us to join in with them." After discussing the offer with Gilfrid, Mike sent a note to Titus Oates, thanking him but declining the offer, explaining that joining forces would make us too large and cumbersome a group.

In late December, Peter Hewett, our legal advisor in Nairobi, informed us that the government had selected our partnership to run Galana. We needed to come in and sign the agreement. Days later, I flew to Kenya and met with Mike and Gilfrid. Pleased and a little bit stunned, the three of us looked at each other, thinking, "What do we do now?"

The deal came together at a complicated time for me. I was busier than ever in my law career and other ventures. I had just been become the lead attorney for the General Telephone and Electronics Company (GTE) in a major antitrust lawsuit in Hawaii. Over the course of a decade, GTE would lose this massive case and enter into a consent decree—a result that would encourage other companies to file antitrust actions and lead to the breakup of the big phone companies.

At the same time, my decade-long partnership in the Heavenly Valley Ski Resort in Lake Tahoe, California, was mushrooming. Hugh Killebrew was energetically promoting a whimsical, publicity-grabbing ski culture that featured swimsuit contests, daredevil skiing stunts, celebrity attractions, and world-class ski racing. As a result, the resort was now a stop on the fledgling professional ski circuit. The World Cup races at Heavenly attracted national television coverage, thanks in part to the charismatic French alpine ski star Jean-Claude Killy. Just as the Galana deal was coming together, Hugh and I were constructing two new chairlifts and planning their inaugural runs, set for New Year's 1968. Once the lifts opened, we would be able to boast that Heavenly Valley was America's largest ski area.

Gilfrid, too, was dealing with complicated issues. His older brother, Charles, had died in a shooting accident, and his father had retired from active management of the family ranching business. Gilfrid was now leading the large enterprise, Kisima Farm Ltd. He was cultivating influential friends in Nairobi and trying to keep the family's prime agricultural lands from being taken for settlement. It was a distinct advantage for us that Gilfrid was rubbing shoulders with people in power, but his new role in the family business had disadvantages. Gilfrid would no longer be available as much as we had hoped, and his family would have to reduce its proposed investment. Gilfrid did not want to jeopardize his family's operations by appearing to control too much of the nation's ranch land.

I was worried that Gilfrid's reduced role in our company would affect our ability to run Galana. Mike and Gilfrid, however, assured me that with a strong manager we could meet our lease obligations, so we hired Charles Moore to take on that role. We also hired Terrence Adamson—brother-in-law of *Born Free* author Joy Adamson—to keep an eye on Galana's existing house and assets until the deal was finalized. To make up for the loss of Gilfrid's investment, I was able to get Alexander "Pug" Atherton, president of the *Honolulu Star-Bulletin*, to invest in our ranch. The government's position as a shareholder, however, remained in flux. Initially, the Kenyatta government planned to hold 47 percent of the company's shares and have at least one member on our board of directors. To limit expenses, however, the government gradually and repeatedly asked to reduce its financial commitment, ultimately settling on 10 percent.

Meanwhile, preparations for our takeover of Galana by 1967 were progressing smoothly. Many observers encouraged us to move onto the property right away, with tractors and graders to draw publicity, and to begin spending thousands of dollars on the operation. Mike, Gilfrid, and I, however, were determined to move slowly and deliberately. We fervently wanted to avoid making mistakes, wasting money, and doing unnecessary harm to Galana's fragile and rare ecosystem.

We began modestly. Mike and Gilfrid cautiously lined up purchases of secondhand tractors, rippers, and graders for roadmaking and hauling machinery. They investigated contractors who could drill boreholes for water and, most importantly, they started to move cattle to Galana from Luoniek and our other ranches. We were using quality Boran cattle breeding stock that Mike and I had produced in our original shareholding. The government was also promising to give us all the support we needed to buy new breeding stock.

Still, it took many months and protracted discussions to iron out the contents of our lease agreement. I was using my best negotiating skills and achieved some concessions, but I could not get the government to agree to extend the forty-five-year lease term. Finally, Frank Charnley, the deputy commissioner of lands, grew irritated at our efforts to change the deal and said, "Take it now or leave it."

We took it.

I soon received the government's Letter of Allotment, dated June 7.[2] It stipulated exactly what Galana Game and Ranching Ltd. was required to do to fulfill the lease agreement. Under its terms, the land was to "be used for the purpose of controlled game cropping, hunting, ranching of domestic livestock, meat processing, the growing of cattle-feed crops, and tourist development only." Above all, the lease required our company to ranch 26,000 head of cattle and manage Galana's natural game population. We were to help it expand to a level that the land could support and protect it against imbalances that might threaten it. In order to prevent overpopulation, the Game Department wanted us to shoot up to three hundred elephant, fifty buffalo, twelve leopards, six lions, and six rhino a year. In addition, we were obliged to bring water to certain areas; build a school, medical dispensary, and administrative facilities; and construct processing facilities to handle the elephant meat that government officials believed would result from the hunting and cropping. After the lease was executed, I would promptly request written permission from Mackenzie to excuse the culling of three hundred elephant. That kind of wholesale killing reminded me too much of the destruction of the buffalo in the American West, an ecological tragedy and a permanent stain on the history of game management in the United States. Mackenzie agreed

But the first task was to approve the deal. I cabled our acceptance immediately. The Galana Ranch was now a reality and nothing—aside from Illie and Christie, who was eighteen and heading off to college in Boston—was more important to me.

2. See appendix, "Letter of Allotment" and "Wildlife Population Analysis."

CHAPTER SIX

WATER WORKS

WE TOOK OVER GALANA OFFICIALLY ON JULY 1, 1967, and plunged into the planning and development of roads, water, and other infrastructure. The reality of building such an enormous enterprise from scratch rapidly proved more demanding than we had expected.

Mike Prettejohn wrote me to report that everything was going well, but he was worried that Galana might take too much time away from his safari business and his goat and sheep ranching at Sangare. Since Gilfrid could not spend as much time on Galana as we had originally hoped, we would also need to recruit a resourceful and skilled on-site manager. "We're going to have an awful lot to discuss when you come out," he told me.

Still, I was brimming with confidence. I wasn't about to let a few practical worries erode my optimism.

I was eager to get to Galana that summer, but Illie and I delayed our trip until late September so we could drop Christie off in Boston for her freshman year. First, though, we made a stopover in San Francisco, where I had to meet with Hugh Killebrew to discuss property purchases at the base of Heavenly Valley. As Illie and I sat on the runway at San Francisco Airport, waiting to take off for Boston, I thought to myself that Tahoe, law, and Kenya made for an interesting life.

Once we landed in Nairobi, though, Galana absorbed my loyalty and attention.

Mike, Gilfrid, and I immediately began to hash out a methodical plan to develop the ranch, based on everything that we had learned over the first three months of grazing cattle, digging boreholes, and dealing with tribal cattle herders. Our most pressing problem was water. We had an enormous grazing area, but we had to get water to it before we could bring cattle there. Cattle need lots of water, and Galana was in equatorial Africa, which got between zero and twenty inches of rain a year.

The most obvious source of water was the Galana River, but there was too little grass there to graze cattle near it. The brush by the river was also filled with tsetse flies, which carry the deadly sleeping-sickness, or trypanosomiasis, parasite. If it weren't for the tsetse problem, Africa would be as famous for cattle ranching as Texas. Because of the tsetse, we had to keep cattle on Galana's grassy plains to the north and dig boreholes to bring water to the herd.

The next issue was to decide where to dig them. If we dug close to the river, we were certain to find water, but we would have the expense of pumping it inland to the prairie. The river's sediment would foul our pumps and probably overwhelm our filters, leading to breakdowns. If we dug our boreholes ten miles out from the river, however, we risked digging a dry hole and wasting precious capital. Still, given the 17,000-foot peaks of Mount Kenya and the 13,000-foot Abedares, there were good odds that we would find rivers of water under Galana. We decided to hedge our bets and do both—dig close to the river and several miles away.

George Classen, an engineer in Kenya's Water Department and an ace at putting water in dry places, believed that we would have to dig 450 feet to find water. Mike's uncle, however, was a diviner. After he surveyed Galana with only a divining stick, he told us that we would find water less than 200 feet below the surface. His predictions of location and depth were amazingly accurate. Supremely rational person that I am, I attribute his success to his long experience and knowledge of topography, rather than to divination.

With George Classen's help, we also built deep ditches to catch rainfall, called *haffir* dams, in Galana's grasslands. To construct a *haffir* dam, we would dig a twenty-foot reservoir and use the excavated soil to build walls on three sides. The walls provided a windbreak that lessened water evaporation from the hot wind. They also provided a bit of entertainment for elephant, who enjoyed sliding down them on their great asses. During dry seasons, when *haffir* dams turned into mud pits, an elephant or two would occasionally get stuck in them and have to be hauled out by tractor.

We also dug channels that carried rainwater into the reservoir from half a mile away. We would then pump the water into a galvanized tank and run it into a trough for cattle.

At first, we worried that elephants might compete for our man-made water sources and ruin them, but we encountered no serious conflict between cattle and elephant. The tanks next to the *haffir* dams were providing water for both. Elephants figured out that to get a drink, all they had to do was drop their trunks over the sixteen-foot-high walls of the tanks. Many rubbed the bottoms of their tusks smooth doing exactly that until we got savvy and capped the tops of the tanks. The heavy-footed pachyderms also helped us by making water pans. Where water filled a depression in the soil, their great padded feet would trample the earth and compress it so it held the water longer. In our early years at Galana, as many as 400 elephants would sometimes congregate at a single water pan.

Generally speaking, things were going well. The Kenyatta government was giving us a great deal of support, and that boosted our confidence. The government paid 60 percent of the cost of constructing water catchments and dams, and Classen was providing invaluable help. We basked in our expanding possibilities. To celebrate, the Galana Ranch threw a big cocktail party in October 1967, at the Muthaiga Country Club in Nairobi, to thank all of the government ministers and others who were helping us with the Galana Scheme. Mackenzie and some of the bigwigs, unfortunately, got stuck in a Parliament meeting and couldn't attend. After the last guests left, our Galana tribe got together for dinner. Parties got pretty raucous in Kenya, and our dinner, Illie wrote in her journal, became "quite wild by dessert . . . buttered red carnations were hurled and flung at the ceiling, where they stuck, very effectively. After dinner, a frantic pillow fight in the bar ensued and ended in a fire extinguisher fight. The Secretary of the Club was stalked and sprayed—the place was a shambles. A great evening."

Our thoughts were filled with plans for the ranch when we returned to Hawaii, but we soon learned that nothing ever happened quickly at Galana. I would set an agenda for what needed to be done and return six months later to find that the project was just getting under way. A year was considered timely. When our Caterpillar tractor broke down early in our first year, for instance, it halted all our efforts to clear a roadway. We had ordered new machines, but heavy rains delayed their delivery for months. The pace of progress was frustrating, but I learned to accept it.

We had exciting news at the end of November, however. Our first 1,140 improved Boran breeding cows and calves, sent by rail from Laikipia to Voi, had arrived at Galana. According to records, they were the first cattle ever brought into the area. We worried, though, about how lions would react to their new neighbors. Gilfrid soon wrote to report that a couple of lions had been nosing around the cattle one night. He thought the lions had gotten a whiff of the cattle when they were chasing oryx, and then stopped to see what they were all about before pushing off. Gilfrid believed that if we could keep chasing the lions away and preventing them from getting a taste of the cattle, we'd be all right. In the beginning that was true, but old lionesses without teeth, along with males that had been kicked out of the pride, quickly discovered that cattle were an easy meal.

A bigger immediate problem was that our first manager was out of his depth when it came to managing cattle and building the herd. We also had to contend with employees who weren't accustomed to our demands for full workdays and workweeks. Despite setbacks and delays, however, we were making progress. When a grader arrived at Galana in early 1968, we began building roads. Indeed, by the time Illie and I returned to Galana in May, the ranch's infrastructure had been improved considerably. We had built two *haffir* dams, each capable of holding 2.5 million gallons of water, and we had contracted with the Water Development Board to build four more. We had completed about seventy miles of new roadways linking existing roads, waterholes, and dams; a sixteen-mile road from the ranch headquarters to the central plains; and various tracks to cattle corrals, or *bomas*. We had finished more than half a dozen airstrips and were planning to build more roadways and airstrips to give us access to the Tiva River in the northwest and the heavy bush country in the northeast. Buildings were going up, too, including a ranch office, an aircraft hangar, and machinery and storage sheds.

Our labor force and payroll were swelling rapidly. In our first year, we had ninety-three employees, not including crews who worked for building contractors or the bearers and trackers we hired for safaris that we ran. Our priority was to employ local people, but out of about ninety or so that we hired, nearly two-thirds quit. As a result, we began bringing in staff from all over Kenya. In part, we were purposeful in hiring people from a variety of regions. With a workforce of unrelated employees from different tribes, we could minimize potential regional and tribal problems. By 1970, the ranch was employing about 140 people from the Kikuyu, Turkana, Kipsigis, Taita, Boran, Samburu, Ndoorobo, and Orma tribes.

Without the Orma, especially, Galana could not have existed. When we took over the lease, the first thing Gilfrid did was negotiate our northern boundary with our Orma neighbors. For a long time, tribal chiefs would not agree to include within our boundaries a particularly attractive piece of land with large, shady *Acacia tortilis* thorn trees. After days of negotiating, the dignified elders graciously conceded, and Gilfrid came to have great respect for the Orma.

Our debt to them went beyond matters of boundaries. The tribe, inheritors of an ancient tradition as herders, were remarkably adept with cattle. The Orma were remnants of the Galla Nation of Ethiopia and northern Kenya, which—after wars with neighboring tribes in the nineteenth century—had migrated south to the rich delta of the Tana River. The Orma depended for their survival on herding cattle, and their distinctive breed of white, long-horned zebu—used as a bride price and slaughtered at weddings and funerals—was considered among the finest in Africa. They also had some of the best native indigenous Boran cattle. We bought selected cows from them, had the animals inspected by a panel from the Boran Cattle Breeding Society, and ran them alongside the Boran breeding herd that Mike and I had started with Boran cows on Luoniek.

It was the Orma, moreover, who came to our rescue after we made our first large purchase of 1,500 head of cattle. We had just gotten them safely to Galana when lions stampeded the whole mob, and the cattle fled in separate groups to the north. Mike and Gilfrid spotted them by plane, and they were well into Orma territory. In the end, it was the Orma who, days later, gathered up our cattle and brought them to our assembly point on the dry bed of the Tiva River.

We were making good progress and had extensive plans. On the drawing board were a manager's house, a dispensary, and a school, as well as a stone causeway over the Galana River. We needed to build the causeway, from the south bank onto our property, not only to meet our lease requirements but also to protect us from the Galana's currents and the yawning jaws of its crocodiles.

UNTIL WE BUILT THE CAUSEWAY, we used a steel cable crossing and then a cable platform pulled by a tractor. The cable was useful, but if the river was very high, it could be dangerous. One day, after we had erected the cable, Mike arrived on the south bank with a new tractor tire that we had to haul across the river. At the time, Terrence Adamson was still working with us. He was in his early sixties, and unknown to any of us, he didn't know how to swim. The Galana was raging, but seemingly without giving it a second thought, Terrence volunteered to walk across the river and carry a cable to Mike, who would attach it to the tire. Terrence tied the cable to his waist and, with a secure belt, hooked himself onto a pulley wheel. He told everyone that if he failed to make it across, they should haul him back. When he reached the center of the river, the cable sagged more than he expected, and he didn't have the strength to pull himself up the bank. Terrence signaled to the men to pull him back, but they pulled too fast. The cable disengaged from the pulley, and Terrence disappeared into the ripping current.

Mike, who had been filming the scene, dropped his camera and ran along the bank looking for Terrence. He saw his hat rush past down the river, but

Construction of causeway across the Galana River.

there was no sign of anyone beneath it. Luckily, Terrence was still attached to the cable and eventually washed up on an island five hundred yards downstream. The men organized a human chain across the river and rescued him.

Clearly the causeway was something we had to build as quickly as possible. Tony Dyer, Gilfrid's brother-in-law and one of our partners in the ranch, had a blaster's license and volunteered to do the rock blasting for its construction. Explosives, detonators, and fuses were strictly controlled by complex legal regulations, and without Tony's help we would have incurred large delays and expenses hiring blasters and engineers.

We planned to build the causeway where the Galana River ran over a shallow rock bed thirty yards wide. After careful planning, we drilled holes in the rock into which Tony could insert sticks of dynamite. Since we didn't have pneumatic hammers to power the drill, we used a more primitive *palunk-a-palunk* system: one man would hold the drill and give it half a turn after another man struck a hammer blow. With this grueling process, a deep hole took a day or more to drill.

In most cases, it would take three days to obtain a blasting permit, procure explosives, and drive them back to the ranch during daylight hours. Tony pushed the schedule, however, by arranging to fly the material directly to Galana, and he procured explosives quickly with the condition that he use them all within a limited time, a requirement that he dutifully met.

On the last day of blasting, when we still had a considerable amount of dynamite left over, Tony decided to get rid of it with a single blast. Unfortunately, a herd of elephant was eight miles from the blasting site when the big bang went off, and the elephants stampeded, almost trampling a nearby hunting party.

We had a safer and quieter celebration when the causeway finally opened in October 1970. Everyone was thrilled to try it out, by foot and by car. Lenita, the wife of our new ranch manager, John Baker, sent us a note happily reporting that Galana was no longer so isolated and that it was much easier to get in and out. It was a big improvement. "Gosh," Lenita wrote, "it is super to know that one is no longer completely cut off!"

CHAPTER SEVEN

CASH COW

CATTLE ARE CASH. The economics of cattle ranching are about that simple. If you buy a dollar bill for ninety cents, you've made a dime on your purchase, and cattle are no different. We made money the day we bought cattle, if we bought well. From recent sales of mature cattle at market, we had a pretty good idea what price our next sale might fetch. We only needed to find good stock to buy at a lower price, and we were on our way to profit.

One of our favorite sources for purchasing cattle was a trader named Saleh Mahdi. Saleh was a third- or fourth-generation coastal Arab whom Gilfrid had first gotten to know when his family owned a farm in Malindi, on the Indian Ocean. We bought thousands of cattle from Saleh at a place called Karawa, north of Malindi, and Gilfrid would do all the negotiating. He and Saleh would sit under a tree for hours as hundreds of cattle walked past them in single file. Eventually, as the two of them came close to agreeing on a price, their tempers would flare. Finally, Saleh Mahdi would grab Gilfrid's chin between his thumb and forefinger, a gesture that meant they were at the final price, and they'd shake hands on the deal.

Saleh had probably never been farther than Malindi, except to buy cattle. But he was always asking Gilfrid—who was then traveling regularly to Britain for wool sales—to take him along on a trip to London. Gilfrid finally agreed. Saleh arrived at the Nairobi airport dressed in a traditional sarong-like *shuka*, wearing a great big army coat over it for warmth. In London, Gilfrid showed him some of the sights and took him to a symphony concert at the Royal Albert Hall. Then one day, when the two of them were strolling down a London street, Saleh suddenly pulled a bank statement out of his coat. He explained that he had wanted to come to London to meet his bank manager. Then he confided that he had been diagnosed with cancer and was terminally ill. He died six months after they returned to Kenya.

After Saleh's death, we had to buy our cattle from Somali traders, some of whom were in dangerous country. The cattle were cheaper and the haggling was less prolonged, but we never had the same bond with them that we had with Saleh. We would buy 2,000 or so head from the Somalis at prices that assured earnings at market. We'd then drive them back to Galana, and we learned from experience not to hold them on the ranch too long. Keeping cattle for longer than we needed to in order to fatten them further for market would hurt our cash flow. The longer cattle remained on the ranch, the greater the chance that they would develop a disease or die. If we kept cattle past their prime, we'd not only have an increase in running costs but we'd also get a lower price for them when we sold them.

Our herds were growing quickly. Our Boran cattle from Laikipia nearly doubled in number in each of our first three years at Galana. By the end of 1968, we had 4,662 head, and we were approaching 8,000 by the end of the next year. The cattle were tended by able and loyal herdsmen from various tribes. Their living conditions were difficult and their days were long. Usually, three herdsmen would look after a herd of 250 cattle. They had to contend with poor water, heat, and malaria and often trekked out to the water hole and back, ten miles a day, under a blazing sun.

Buying cattle in southern Somalia, 100 miles north of Galana Ranch.

Our herders also often had to deal with lions and other predators that threatened the cattle. We would greet them, saying, *"Habari yaku?"* (How's it going?), and they'd typically reply, happily, *"Habari mzuri sana."* (Everything is very fine.) But one beat later, they'd add, *"Lakini . . ."* (However . . .) and inform us that a lion had just killed a milk cow, or give us another unfortunate piece of news.

In general, we were proud of our herdsmen and admired them greatly for their stamina and intelligence in caring for the animals. We realized, though, that they could be more efficient. We noticed, for instance, that after they released the cattle at seven A.M. from their *bomas*, the herdsmen often lingered to have long breakfasts of milk and porridge, or *posha*. That meant that the cattle were spending a substantial part of their feeding time where there was little grass left, when it was important for the herders to push the cattle out onto the grasslands. We also had to caution them not to bring all their herds to a watering pan at the same time while they took their *chai* (tea) breaks and conversed. Five hundred or more cattle drinking from the same water pan at the same time was not good for the water pan or the cattle.

We took a good deal of the herders' advice, though. They informed us, for example, that when lions approached the herds at night, the cattle would break out of the tight *bomas* and spread out. So we stopped using the enclosed *bomas* and let the herdsmen keep the herds together at night around campfires. If lions came near, the herdsman would call the cattle, comfort them, and let them mill around close to the safety of the fires.

We were acquiring a firm grasp on the economics of cattle ranching in Kenya, but in our early years we knew that Galana's elephant and wildlife were our greatest asset. When we started, Galana had some 6,000 elephant, 300 rhino, and a multitude of other wildlife. The game soon attracted an international clientele of wealthy hunters, who paid for hunting licenses with fees tailored to each species. We discouraged a shoot-anything-that-moves mentality and wanted to attract hunters who would pay large sums to shoot fewer animals and be guaranteed the availability of larger ivory or horn. We ensured that elephants were mature, past breeding age, before they could be shot on Galana and that old bulls were shot away from the breeding herds. In our first year, nine visiting hunters shot five elephant, seven lesser kudu, six zebra, three gerenuk, two warthog, eight oryx, eight Grant gazelle, one waterbuck, eleven eland, six dik-dik, and one buffalo. In 1970, thirteen overseas sportsmen hunting on Galana shot seventeen elephant, with tusks weighing an average of almost seventy-five pounds. As a result of our hunting business, the government of Kenya reaped important foreign exchange, and the ranch had revenues that allowed it to grow at a rate that would have otherwise been unattainable.

To accommodate our hunters, we constructed a handsome lodge on the Galana River. Built of local stone and roofed with *makuti* thatching, it had a dining lounge and three comfortable sleeping *bandas* with bathrooms. When Mike was sidelined from leading our safaris for a while, we hired our first full-time white hunter, Barry Roberts, to run our game operation.

Given the ranch's isolation from the rest of the world and its harsh environment, people occasionally got on each other's nerves—especially when they were in the competing ranch and game sides of the business. So, to avoid problems, we built Barry a small house upriver from our headquarters and away from the home of our managers, Charles and Jane Moore. What we didn't realize was that we had built Barry's house in the middle of a rhino path.

On his third night in his new home, he was awakened by a steady thumping on the wall, which was constructed of pressed mud and wattles. At first, Barry thought an earthquake was shaking the house, but the pounding continued and seemed to come from a corner of the room where an ammunition box, which held his valuables, was propped on poles. Barry finally roused himself out of bed to investigate—just as the head of a rhinoceros bashed right through the wall. Fortunately, the ammunition box crashed onto the

Main ranch house on Galana River, 1972.

head of the surprised animal, which withdrew and continued, seemingly unperturbed, along its way.

Unlike the rhino, most of our game animals—especially the elephant—would go into a near panic, in our first years, when they got a whiff of a man or heard the sound of vehicle or aircraft. The reason was mainly the earlier excessive cropping of animals by the Game Department. When we took on the Galana lease, the government wanted us to continue shooting three hundred elephant a year—not only to control the size of the herds, but as a source of meat to be sold to citizens. We never thought this was the right or practical thing to do. In fact, in our earliest meetings, I told Bruce Mackenzie that we would have nothing to do with that sort of wholesale culling. It was simply wrong. Besides, there was no way that after killing an elephant fifty miles out in the bush, we could transport its meat to an abattoir quickly enough to be processed and canned for human consumption. It was impossible to do it in a way that ensured human health and safety. It would have been legal and easy for us to finance our cattle operations by cropping elephant for ivory. After we took over Galana, however, we did not do any elephant cropping and we allowed hunters to take off only a few old bulls a year, for a substantial fee.

Within a couple of years, our decision to refrain from cropping was showing results. Our game, particularly the elephant, was calming down, and most no longer bolted at the first sign of human, vehicle, or aircraft. The elephant began to return to Galana's water holes, although they continued to keep their distance in the daytime from *haffir* tanks and the watering places that the cattle used.

A portion of the hunters' fees went to the government, but other revenues from our game operation helped us support a managed game environment and protect the wildlife against poaching. Beginning in the late 1960s, when poaching was an escalating problem, we organized our first anti-poaching unit, headed by Barry Roberts. The patrol included six trackers accompanied by two Game Department scouts, who could authorize arrests and give persuasive testimony in court. After the team discovered several freshly killed elephant and giraffe, as well as skins of lesser kudu and other small game, they arrested a number of Waliangulu poachers, removed their traps, and confiscated their poisoned bows and arrows. Seven of the poachers were later convicted.

By 1970, our safari business and game management efforts were taking hold. Major construction at Galana was ending, and expenses were drastically curtailed. When Illie and I arrived that August for our third stay in twelve months, I looked down from the Cessna 180 I was piloting and saw the impressive changes that had taken place. Our new headquarters building was a

Typical Galana Ranch houses, 1970s.

handsome structure. We had completed the hunting lodge, the causeway, and housing for staff and labor. We had built 105 miles of roadway and firebreaks, a 10-mile pipeline, water holes and dams, game pens, yards, and a game lookout that could accommodate twelve viewers. We had more than 7,000 head of cattle. With the construction of a big dam and help from a bank loan, we were projecting that Galana would soon support 10,000 head. I felt confidence and pride, and the anxiety I had felt in previous years over the cost of development began to fade.

Time swept by fast on that visit. My notes are scant; there was too much to do to lose time writing in a journal. Galana had become a way of life for us, and it no longer seemed necessary to write down everything that happened. After all, I reasoned, we would be living these memories for forty-five years. The end of our time at Galana, the year 2012, was a lifetime away.

CHAPTER EIGHT

DOMESTICATING SCIENCE

"MARTY, IF YOU EVER BUY RANCH PROPERTY up here or anywhere else, don't make long-range plans for your first ten years. You've got to really get to know the land first," an old rancher in British Columbia had advised me years earlier.

His voice played in my mind as Mike Prettejohn and I sat around a campfire one night, discussing how we would address our lease obligation to use wildlife as a source of protein for Kenya's skyrocketing population. Bushmeat had, of course, long been an important source of food in parts of Africa, and domesticating game animals—penning and herding them on the open plains without shooting them—seemed the most promising option to us.

Since 1935, research on breeding tamed eland had been under way at the Soviet Union Institute of Acclimatization and Crossbreeding in southern Ukraine, although the goal there was to use eland as a dairy animal. By the 1960s, there was mounting interest worldwide in game ranching to replace or supplement cattle farming on marginal lands. Scientists noted that an estimated 31 percent of the earth's surface is either arid or semi-arid terrain, with an average rainfall inadequate for continuous crop production. At least half of this land produced a large range and quantity of plants. Feeding these plants to animals was the only way to convert them into food for humans. Biologists, ecologists, and conservationists argued that in Africa, game ranching would be ecologically beneficial and economically more profitable than cattle ranching. Sir Fraser Darling, an ecologist and conservationist-philosopher, considered it folly to import domesticated animals to Africa when the region possessed ample indigenous hoofed animals that were already adjusted to the land's resources. As he put it, "To exchange the wide spectrum of twenty or thirty hoofed animals living in delicate adjustment to their habitat for the narrowed spectrum of three ungulates exotic to Africa—cattle, sheep and goats—was to throw away a bountiful resource." The greatest early proponents of game ranching included the eminent British environmental biologist E. Barton Worthington and American ecologist Raymond F. Dassman. Dassman's pioneering work on African game ranching fostered the field of ecodevelopment and helped make ecotourism a household word.[1]

Still, game ranching had been developed on a significant scale only in southern Africa countries, below the Zambesi River, and efforts in Kenya were limited. In 1967, David Hopcraft had launched a 20,000-acre experimental game ranch outside Nairobi, on the Athi River. The research at Hopcraft's ranch, underwritten by the U.S. National Science Foundation, spawned great interest and considerable publicity, but it involved shooting impala at night rather than domestication. Another property, the 1,256-acre Mount Kenya Game Ranch outside Nairobi, was developed to breed eland as a commercial source of meat. Actor William Holden became a partner in the venture in the late 1960s. Our project, however, would be of a different magnitude, with greater scientific discipline and order, than any of these schemes.

Barry Roberts and I had occasionally chased down some eland and oryx to see how difficult it would be to capture them for domestication. After those first attempts, Mike Prettejohn queried the Kenya Game Department about the feasibility of exploring wildlife domestication at Galana. Game officials put us in touch with John King, a Cambridge University PhD and veterinarian officer for the department. John had come to Kenya six years earlier while working on his doctoral dissertation on equine reproduction, and he had studied adaptive characteristics that wild ungulates use to exploit African ecozones. In the process of working on that study, he had developed some expertise in capturing the animals and using tranquilizing medications. John was interested in our

---

1. H. P. Ledger, along with R. Sachs and N. S. Smith, wrote "Wildlife and Food Production," published in *World Review of Animal Production* 3:13–37. In 1983, Ledger wrote "The Rational Use of Wild Animals," in *Domestication, Conservation, and Use of Animal Resource,* ed. L. Peel and D. E. Tribe, vol. A1 of *World Animal Science* (Amsterdam: Elsevier Science Publishers). For this chapter, I also closely re-read the research articles prepared by Brian Heath in collaboration with John M. King and others for the Galana Game Farm Research Project, for the African Wildlife Leadership Foundation of Washington, D.C., including "Game Domestication for Animal Production in Kenya: Theory and Practice," first published by World Animal Review (Rome: Food and Agricultural Organization of the United Nations, 1975), no. 16, and revised in 1977 for the *Journal of Agricultural Science* 89 (1977): 445–457.

proposal, and in February 1970 he began work on a game farm research pilot project at Galana.

At first, we considered domesticating buffalo because they look and act a lot like cows and move in large, closely packed herds. Eland was another candidate for domestication, since earlier efforts in Tanzania and Rhodesia suggested that they were docile animals that could be run like cattle. Last were fringe-eared oryx, which, like the other game, were common on Galana. Oryx, however, had long, needle-like horns, and no one had made any efforts to tame them. Lions avoided attacking oryx, and they were believed to be too aggressive to domesticate. Our project aimed at evaluating the productivity of each of these species separately and in combination with cattle and sheep. Our ultimate objective was to obtain the maximum yield of meat per acre without destroying Galana's fragile ecological balance.

Our research project got under way when Mike recruited Wing Commander M. W. "Punch" Bearcroft, head of the Kenya Police Airwing, to fly down from Wilson Airport in Nairobi with a government helicopter to help us round up some animals into a plastic *boma*. Mike and Punch were old comrades who had worked together on numerous sorties during the Mau Mau rebellion. Punch had lost his right hand in a motorbike accident as a boy, but he flew the helicopter ably by strapping his right arm to the joystick with a bit of rubber from a car's inner tube. John King directed the operations from the ground.

I had proposed and put together an $11,000 budget for John's research, with contributions from Galana Game and Ranching, the African Wildlife Leadership Foundation (AWLF), and, most significantly, Texas A&M's Caesar Kleberg Research Program in Wildlife Ecology. By the time Illie and I arrived in August 1970, the AWLF and Texas A&M researchers had already agreed to participate in the pilot study. John Rhea, AWLF's executive director, and Frank Minot, its director of African operations, believed that although our preliminary work would only establish a principle, it could make a significant contribution to wildlife conservation.

They pledged their organization's support, but they emphatically cautioned me to tamp down criticism on the ranch of John King and his research methods. To the rough-and-tumble Kenya cowboys at Galana, John was a "boffin"— British slang for a scientist or egghead—and his imperious manner had earned him the moniker "King John." His style rubbed them the wrong way, and they wondered if his scientific research bore any useful relationship to the realities of ranching. John, in turn, came to question Mike and Galana's commitment to serious research. I promised I would end this conflict, and I did. The fact is, John was a vital participant in the project. The scientific progress we made at Galana owed much to his dedication, his fund-raising efforts, and his skill at selecting and guiding a corps of talented young scientists.

His first excellent hire, as project manager, was a twenty-year-old Kenyan named Brian Heath. Raised on a farm on the slopes of the Rift Valley, Brian had attended high school in Nairobi. He had distinguished himself there by winning an award for his original work preparing a museum-quality bird collection of 400 species. After he completed his secondary education, Brian landed a job working as an assistant to *Born Free* author Joy Adamson. For nine months, he had looked after her pet cheetah while she recuperated in England from a hand injury that she'd suffered in a car accident. When John hired him, he was grateful to come to the still-primitive world of Galana. He was little more than a teenager, but he would prove to be one of the most astute, productive, dependable, and durable members of the Galana tribe.

Brian could be quiet, to the extent of not talking for several days on end, but he was a fount of knowledge. He also had a unique capacity for managing the diverse herdsmen and a raft of sometimes wild-spirited young scientists, who called him "Brain" or "Bwana Brown." He was ever helpful to the scientific neophytes who came to the ranch to investigate wildlife capture and domestication, resistance to drought and diseases, productivity of the animals and the vegetation, and the marketability of meat from wildlife as a source of protein. Many of those young scientists went on to make their marks as world-class conservation experts and credited Brian as a coauthor of their Galana research.

We caught the founder stocks for our project—three buffalo, a few eland, and a few oryx—on the ranch or brought them in from other areas, and Daphne Sheldrick sent us some buffalo orphans from Tsavo East National Park. We started by experimenting with a variety of capture techniques—using nets, spotlights at night, and tranquilizer darts and driving herds into a fabric-walled corral. Gradually, we honed our methods. Since buffalo are not generally found in the open, it proved easier to dart them with immobilizing drugs from a car or helicopter than to chase them to exhaustion. Brian found, however, that we could chase eland to a standstill in a few minutes in open country. We caught only eland that had been weaned, since unweaned animals had to be bottle- or hand-fed, which was too labor-intensive.

We had our most striking success in capturing oryx. Brian discovered that if we cut one oryx of the right size from the herd and chased it down with a Land

Exhausted oryx captured after a rundown on the ranch.

Cruiser, it took less than five minutes before the exhausted animal would turn to bay and flourish its long horns threateningly at the car. That was the signal for the herders riding on the back of the vehicle to leap down and restrain the oryx by the horns.

A second method involved chasing a herd of oryx toward a trap—a Hessian curtain lying flat on the ground. The herd would sometimes sense something amiss and veer away before getting close to the trap. The drivers of the Land Cruisers, however, got pretty adept at anticipating the herd's movement and could drive the lead oryx back to the trap area. The rest of the herd would follow, and when they got to the trap, ten to twenty herdsmen would spring from the vehicles, raise the curtain from the ground, and secure the capture. The oryx would then be driven into a sorting chute. We selected only those that were in excellent health and two years old, the optimal age for breeding. We let the rest of the animals go back to the wild.

Capturing oryx was exciting stuff. Zoologist Marlin Perkins brought his cameras to Galana in the mid-1970s to film the ranch for his famed *Mutual of Omaha's Wild Kingdom* television program. Perkins was by then a familiar white-haired, avuncular presence on the popular TV show, which he had started when he was director of Chicago's Lincoln Park Zoo. Our oryx capture routine was the star segment of the Galana show, and Perkins enthusiastically

narrated each step of the process. "The careful, humane capture of large herbivores is masterfully accomplished," he informed his audience, and he called
the ranch "the Galana Experiment" in dramatic cadence. It was, he said, "an
example of excellent land use" and "good conservation."

For the doctoral students who pitched in to help Brian, the adventure of
capturing oryx was an exciting break. A lot of the science they did was just
a hard slog. Ecologist and rural development specialist Jeffrey Lewis, for example, had to observe animals nonstop for twelve hours while he walked in the
bush. Every four minutes, he had to record how many animals were doing what
type of activity in what type of vegetation, noting the temperature, solar radia-
tion, wind speed, and soil type at each location. For Jeffrey, the cowboy work of
oryx roundups was a welcome and exhilarating change of pace.

To the scientists, I must have seemed like a bit of a romantic—arriving twice
a year on my vacations from the law, full of enthusiasm to go out and capture
oryx, while they lived a gritty life on the ranch, doing their often tedious, full-
time experimental work. To them, Gilfrid was a sheep and cattle man, with a
vested interest in maintaining the dominance of cattle ranching. Mike, whose
job was to oversee our hunting operation, must have seemed somewhat indifferent to their painstaking, data-collecting efforts. But in truth, we were anything but indifferent, and with the cooperation of the Kenya government, our
domestication project made remarkable progress.

After we captured game animals, we would blindfold those we'd selected
for domestication. Once they were calm enough, we would truss their forelegs
and hind legs and put them into pens. The enclosures were lined with grass to
prevent injuries if anxious animals hurled themselves into the walls, and the
darkness of the grass-lined cells had a calming effect. We fed the game animals
hay, local grasses, and protein-rich lucerne (alfalfa) and gradually removed the
grass padding from the pens after the first week. Once the animals were calm
enough, we let them out onto a verandah, and they were able to retreat to their
pens if they were frightened. At first they had little contact with humans, but
over time, as they grew accustomed to people, we gave them more freedom and
moved them into progressively larger enclosures. When they had fully settled
down, we moved them into a hundred-acre paddock, where they spent three
to four months before they were absorbed into the main herds of each species.
The game animals were herded out to graze during the day and brought back
to a corral at night. There was no fencing to prevent escape, because domesticated animals, by definition, stayed with their herdsmen. John King and Brian
Heath documented the process in their paper "Game Domestication for Ani-

mal Production in Africa," which was published by the United Nations' *World Animal Review.*

Each species, we learned, reacted differently to captivity. We made a brief attempt to domesticate gazelles, but it failed quickly; regardless of the size of the enclosure, they would hurl themselves at its walls and injure themselves. Buffalo, on the other hand, did not need to be penned at night and slept peacefully around the herders' campfires. Domesticated cows, particularly those with calves, became very placid. Castrated males, however, could be boisterous, and bulls could be dangerously aggressive as they grew older. After we had studied buffalo for a time, we dropped them from the program because of their bovine feeding habits and temperament.

Eland, we found, were easily tamed and would take food from the hand within a short period of time. We were also able to accelerate the taming process by keeping domesticated eland with newly captured animals. But eland were nomadic, selective browsers, and harder to herd than grazing animals. They spread out and moved quickly while they were feeding. Many domesticated eland would go off for up to three months at a time in search of better food. They then suddenly would reappear at a water trough and were reabsorbed into the herd. The fierce heat of the African sun was also a problem

Illie with domesticated eland.

for the eland. In the wild, they feed at night or in the shade to control their temperature. But domestic eland had to feed and be active during daylight hours, because we penned them at night, and their exposure to the sun was unavoidable.

We eventually focused our efforts on domesticating oryx. On their first day of captivity, oryx charged around their pens, but they settled down quickly and usually joined the domesticated herd within six weeks. Unlike the eland, escaped oryx seldom moved more than a couple of miles from the pens and were relatively easy to recapture. Few oryx calves born in captivity on the ranch ever ran away. Most importantly, given Galana's semi-arid environment, oryx were able to gain weight by grazing on less vegetation than cattle needed to maintain their weight. They also needed only a quarter of the drinking water that Boran cattle needed and half the water required by Dorper sheep, a breed that was well suited to arid areas.

Each study the scientists undertook contributed another element to our understanding. Wildlife biologist Tim Wacher published the first detailed study[2] on the ecology and social organization of the fringe-eared oryx. Jeffrey Lewis, who was funded by Texas A&M's Kleberg research program, explored how the process of domestication affected the apparent ecological advantages of eland, oryx, and buffalo over cattle, sheep, and goats. He also studied the effects of their enclosure at night to protect them from predators. Lewis found that the oryx were much less compromised than eland or buffalo, because of the way they fed and their superior capacity to resist heat stress. According to Oxford-educated Mark Stanley Price—now chief executive of the Durrell Wildlife Conservation Trust—the oryx was the one mammal that allowed its body temperature to rise a few degrees during the day, when it was exposed to high air temperatures. Price, who worked on Galana from 1974 to 1978, found that the oryx dissipated the heat at night, ensuring that its delicate brain tissue cooled sufficiently. We discovered that the oryx needed less water because they produced small quantities of urine, very dry feces, and small amounts of rich milk. They had fewer sweat glands than the other animals and cooled themselves by panting. The studies also indicated the extent and speed with which grazing oryx could be moved, the length of confinement they needed, and the amount of privacy that cows required when they were calving.

2. Timothy John Wacher, "The Ecology and Social Organization of the Fringe Eared Oryx, on the Galana Ranch, Kenya" (Ph.D. diss., St. Edmond Hall, Oxford, and Animal Ecology Research Group, Department of Zoology, University of Oxford, 1986).

Secluded lorry for transport of captured oryx to holding area.

First stage of oryx domestication—transferring captured yearlings to isolation hut.

Interestingly, an important element of the game domestication process is the quid pro quo—the benefit the animals derive from herders in exchange for their captivity. Herdsmen supplied a sense of security that was of obvious value to the oryx. Individuals in wild herds frequently raised their heads and looked for danger—interrupting their feeding and reducing the amount of food they could consume. Galana's domestic oryx, by contrast, kept their heads down and fed continuously. Often, it seemed they understood that a herdsman standing nearby with his spear was there to protect them against predators. After spending hours watching the interactions of Galana's domesticated oryx and their herders, I was convinced that the animals had not surrendered their freedom in the domestication process. Instead, they had rather adapted to man as a protective factor. The oryx, I sensed, were habituated, not domesticated; they still had a wild look deep in their eyes.

After five years, we concluded that only the oryx were well enough suited to conditions of heat and dryness, in their physiology and behavior, to have serious potential for large-scale domestication at Galana. After we had built up several herds, we sent some oryx meat to a butchery in Malindi to see how it would sell. There was considerable interest. The oryx meat sold for twice the price of beef and was well received in hotel restaurants by tourists who were

Domesticated oryx with full-time herder.

interested in savoring exotic game meat. We then applied to the government for a permit to increase the scale of capture, enlarge our herd of oryx to four hundred head, and sell the meat on a commercial basis. The government turned us down, believing that approving our application would increase the sale of illegally poached animals.

We were never certain that we could have produced oryx meat in a profitable way. We knew that we had to have at least four hundred oryx to sufficiently spread the costs of overhead. Even then, it was doubtful—given the labor-intensive costs of raising the animals—that oryx would ever have been more than a supplement to cattle ranching. Nevertheless, despite the government's rejection of our request, we could claim a great deal of success. Most significantly, we established that the domestication of game animals made sense only on marginal lands, such as in southern Sudan, that are unsuitable for cattle because of insufficient drinking water or annual rainfall. Game ranching is not likely to be used anytime soon in most parts of the world—despite increasing global needs for protein—because there are still plenty of grasslands for grazing cattle. In places where the economics are right or when there is too little cattle-producing land to meet the food needs of the population, however, oryx could be put to productive use. We also established a domestication technique that could be repeated in different areas, with different species. We provided another reason to preserve oryx in the future, and we demonstrated the economic contribution that the animal might make.

At the very least, the research we did at Galana is there for the future.

CHAPTER NINE

THE END OF INNOCENCE

BY 1972, I WAS MORE OPTIMISTIC THAN EVER about our ability to keep investing in Galana—both financially and as a way of life. Illie and I had moved from Honolulu to San Francisco, where I had opened a branch of Goodsill, Anderson, Quinn & Stifel to handle Hawaii-related work from mainland corporations. At the same time, I had become a partner in the respected San Francisco law firm of McCutchen, Doyle, Brown & Enersen. At Galana, too, our enterprise was advancing well. Our safari, game, and cattle ranching operations were taking hold and meeting the projections we had established. Our unique, diversified, forward-looking business was attracting increasing attention from the press, the ranching and research communities, and a growing clientele of hunter-conservationists including Bob Kleberg, president of the global King Ranch empire that was based in South Texas. Kleberg was so excited about our game ranching that he asked us to supply some of our domesticated oryx herd to the King Ranch. Ultimately, however, that proved impractical. Under U.S. law, the first generation of oryx would have had to go into a zoo, as a precaution against dangerous African ticks and diseases. Only the second generation would have been permitted to go the King Ranch.

Our cattle operation was now close to 10,000 head—still 16,000 shy of our long-term goal of 26,000 cattle, but well ahead of what our lease required for our first five years. The mortality rates for our cattle were slightly higher than our 2 percent goal, and the birth rates were slightly lower than the 80 percent we wanted, but they were moving in the right direction. We planned to expand our cattle herd to 18,000 head over the next five years and had sufficient confidence to negotiate a $200,000 loan from the World Bank for cattle purchases, water development, and operating costs.

After some management turnover, we hired J. H. Patterson to run our cattle business, and at the end of 1972 we brought on Ken Clark as our new resident hunter. Ken had previously been a hunter for Ker, Downey & Selby Safaris, the oldest and most distinguished safari firm in Africa. With revenues from our cattle, hunting, and game enterprises growing steadily, we were looking forward to our first profitable year.

We had one nettlesome problem. The Orma—the tall, handsome, semi-nomadic people who were our northern neighbors—were regularly moving their cattle across the boundary that divided us to graze on Galana's grasslands. To settle the issue, we held a series of successful meetings with the Orma elders and agreed, as a concession, to dig and maintain a borehole for them on our northern boundary. We never completely resolved the problem, but it never erupted into a serious conflict. Usually, if Orma herders were trespassing up north, we would tell them to go, and they would do so for a couple of days. We remained flexible and used polite pressure, understanding that when herdsmen were out in the middle of the bush, they were not necessarily aware that they had come onto our property. The truth is, when you have a lot of grass and you're not using it, it's pretty hard to tell an African who needs it to keep his cattle off it. More to the point, we wanted to do what we could to accommodate the Orma, who had done so much to help us.

In general, I didn't want to have a heavy-handed presence in this vast, wild country. By the early 1970s, I was growing concerned about protecting the best of primitive Galana. I didn't want to build more permanent structures that would mar its mesmerizing empty landscape. So when our lease rent rate came up for review, I proposed—and the government agreed—that we limit cattle ranching to just 500,000 acres, leaving Galana's other million acres free for wildlife.

About the same time, I insisted that visiting hunters reserve a minimum three-week safari. I didn't want hunters at the ranch who would shoot at the first animal they saw. I wanted patient hunters who knew they had three weeks to understand our red-soiled land and riverine habitats and to track, select, and earn a proper, mature trophy. I hoped that that strategy would allow us to acquire enough from hunting to benefit the ranch and employ locals as trackers and safari staff, while minimizing pressure on our game. As it was, we continued to keep well below the Kenya Game Department quotas we were allotted, especially for elephant. The numbers of animals taken off for hunting had virtually no impact on Galana's game population.

Personally, I found one of my greatest joys in the hunting that I did on Galana. It could never be equaled or duplicated. In fact, after I bagged a big bull elephant in 1971—whose tusks each weighed 116 pounds and measured 9.5 feet on the curl—I was finished with elephant hunting and did no more shoot-

Marty with trophy tusks, at Yalou River, 1974.

ing except for lion control. I was ready to stop, and there were plenty of other things at Galana to keep me busy.

On the ranch, Illie and I worked together, and I was in charge of the long-term vision and longer to-do lists. One of them read:

Garbage: Dig pit for burying

Kitchen: Repair fresh meat storage; screen broken, flies inside

Bandas: Repair mosquito nets; brush and clean showerheads; remove water stains around wash bowls; mend and replace sheets when holes develop

Illie and I were a well-drilled team. She also checked religiously to make sure that I had enough fuel in the car before I drove anywhere. That was no small matter. The ranch now had hundreds of miles of roads, but once you were out on them, there was nowhere to get gasoline if you ran out. A wrong turn or two—easy enough to make in that flat terrain, with little to distinguish one place from another—could land a person in deep trouble. In fact, after the causeway was opened, we once came very close to losing a Japanese couple who drove onto our vast property to do some sightseeing. We caught them just in time. They could easily have gotten lost, run out of gas, walked miles for help where there was none, and perished under the punishing sun. After that incident, we posted a warning sign at the entrance to the causeway.

Illie spent hours painting almost every day, and she loved seeing wildlife in the bush. Amazingly, she was always meticulously turned out, even after five hours on Galana's bumpy, dusty roads. She never seemed to have a hair out of place or a lipstick smudge, though the rest of us would be sweaty, grubby, and gasping for a beer. Stepping out of the grimy vehicle, she often looked like she had just left a Manhattan beauty parlor. "Goodness knows how she did it, but it was very impressive," remembered Jeffrey Lewis, who grew accustomed to hearing me call her affectionately by the one-word name "Illiedear."

There was much work to do on Galana, but the isolation of the place inspired raucous fun as well. Each year we held a rodeo, including a "weaner"-(young steer-) throwing contest, and the brainy young wildlife biologists were especially fond of having an occasional elephant dung fight in the Galana River. It was a hazardous place to play, given the number of hungry crocodiles that croaked and snapped their jaws along its banks and slithered silently beneath its murky waters. On one occasion, when it was raining and Chris Flatt, our ranch administrator, expected major flooding, he, Barney Gasston, and two others decided to try their luck at shooting rapids on the river. The water was brown, with huge ridges of foaming waves and whole uprooted trees hurtling downriver at great speed. The four men set off in a rubber raft—paddling, laughing, drinking, and more than a little anxious. They spotted what at first looked like a hippo swimming toward them from the bank, but it was actually a monster of a crocodile. The croc no doubt mistook the raft for a dead animal floating on the flood, and it came right at them. When it was within several yards of the men—who had a millimeter of rubber fabric between their backsides and its huge teeth—the croc disappeared under the water, and the men, holding their breath, sped away from it down the river.

Given Galana's isolation, Angela Sutton, who ran the ranch office in Nairobi, was a vital link to the world. Sooner or later, just about everyone was dependent on her for some supplies, from transmissions to toothpaste, that could, after a while at Galana, seem perfectly exotic. But for the most part—with the exception of Illie and the wives of our managers—we were a male tribe. Without a doctor, Galana was no place for women, especially a pregnant one, or small children. The herdsmen who worked for us would generally stay at Galana for a few months, then go home to their wives and families.

Six of our herdsmen were designated as "sergeant majors"—a term of respect we brought over from the Game Department. They supervised the rest of the group, and one of them, Jomo, was our majordomo. He had come to us through Gilfrid—who, after catching Jomo poaching on his upcountry land, had given him the choice of going to jail or coming to work for us on Galana. Jomo chose Galana. He learned quickly, worked hard, knew how to supervise others, and became our most dependable employee. Sometimes we asked him to teach the young men who came to work on the ranch for a period of time, and he patiently and tactfully educated them all about livestock on the Galana grasslands.

It was an environment, in the early and mid-1970s, that was under stress. Sub-standard rainfalls, year after year from 1971 through 1975, took a devastating toll on the entire region. Galana had ample water resources, as did Tsavo National Park, but the drought had a catastrophic effect on the elephant population. The tuskers were actually dying of malnutrition rather than thirst. Before the drought, Tsavo had struggled with an overpopulation of elephants, a problem that had first been encountered in places such as Murchison Falls National Park in Tanzania. Growing herds were compressed into the park's safe haven by human encroachment in surrounding areas, and their browsing was transforming the thick woodlands into open grassland. At Murchison Falls, the environmental destruction had caused the disappearance of the chimpanzee, the forest hog, and much of the park's bird life.

A prominent scientist named Richard Laws determined that culling elephants was the only way to restore the balance between elephants and their habitat in the park. As a result, between 1965 and 1967, some 2,000 elephants out of 14,000 in Murchison Falls were shot. Tsavo, Kenya's most famous wildlife sanctuary, was confronting the same problem, and its founding warden, David Sheldrick, commissioned Laws to study the situation there. In 1969 Laws recommended cropping 3,000 of Tsavo's elephants. Sheldrick—who had devoted twenty years to developing and protecting the park and had once been a proponent of culling—vetoed the plan. I agreed with his objections to culling. Sheldrick had come to believe that it was preferable not to interfere with the elephant population, even if it meant dooming many animals to starvation. He believed that nature would kill the weaker ones and that those that survived would produce a stronger population better able to deal with the next drought or habitat change. The government backed Sheldrick, and Laws resigned.

The drought of the early 1970s made the controversy more acute. There wasn't enough food for the elephant herds, and the parched vegetation that did exist had extremely low levels of protein. Other herbivores, with more efficient digestive systems, could extract enough protein from the available food to survive, but elephant and rhino lose a lot of protein in their waste. Young elephants, not yet adept at foraging and too small to browse on higher trees and shrubs, were the first to weaken. They slowed the cows down, who quickly declined. Malnourished, with compromised immune systems, the cow herds lacked the strength to leave permanent water sources in search of food.

Ultimately, some 10,000 elephants, a quarter of the population at Tsavo, died of malnutrition because of the drought. Some of them sought relief by crossing the nonexistent boundary into Galana, and we found many small, dead elephants. We made numerous attempts to rescue those that had been abandoned and to pull a few larger cows out of water pans, where they were stuck, but their condition was generally too dire for them to survive. The die-off of elephants intensified the culling debate, with each side claiming that the results proved its argument. Pro-cullers insisted that cropping would have resulted in lower overall mortality, while anti-cullers claimed that the drought proved that natural events control populations.

Amid the sadness and these challenges, there were other tragedies. In August 1973, Gill Prettejohn—Mike's wife and the mother of their three children—died after an accident. Her death was devastating to Mike, his family, and the rest of us at Galana. Mike had to pull back from his responsibilities at the ranch to tend to his personal affairs. Gilfrid stepped in to provide more supervision of our cattle operation, and I became more actively involved with the game enterprise.

Three months later, we had another jolt. The government of Kenya enacted an eight-month ban on elephant hunting in response to a public outcry over the Tsavo die-off. The ban was also provoked by alarm at the reports of increased elephant poaching in Kenya and the rest of Africa. The price of ivory paid by illegal buyers in the bush had increased dramatically; it was now double the price that Kenya's Game Department paid for legal ivory. Since 1971, gangs of poachers, armed with automatic weapons, had moved into Tsavo, and the park's elephants—concentrated along the river and near permanent water during the drought—were easy targets. There was growing evidence that the poaching epidemic was being fueled by corruption in the Game Department and at high levels of government. As greed and patronage crushed the cooperative, forward-looking spirit of *harambee*, elephant tusks and rhino horns became vehicles for laundering money and taking it out of the country.

Since Kenya's independence, Jomo Kenyatta had been consolidating and expanding his executive power. In June 1964, six months after he took office as prime minister, the country had become a republic, and Kenyatta served as its first president. In 1966 he was elected, unopposed, to a second term. He moved quickly to broaden his constitutional powers and place members of his Kikuyu tribe in the highest levels of government. In 1969, his popular minister of economic planning and development, Tom Mboya, suggested in Parliament that a number of Kikuyu politicians were enriching themselves at the expense of other tribal groups. Speech that could not be gagged in Parliament could be silenced on the street, and Mboya was assassinated on July 5, 1969, outside a Nairobi pharmacy. After his death, Kenya's nascent nation-building enterprise began to fall apart, and the optimism inspired by independence began to fade. Ethnic patronage became entrenched, and there were rumors that the illegal ivory trade was enriching government ministers, senior officials, wildlife authorities, and members of Kenyatta's immediate family.

In 1973, the black market price for ivory jumped from $50 a kilo to $700 a kilo, and the number of elephants slaughtered in Kenya leaped to a thousand a month. Officially, the government condemned poaching. On Galana, after we made numerous requests, the leader of our anti-poaching unit, Waiwha Kanimi, was finally given the power to make arrests. We supplied him with a W. W. Greener shotgun and a Toyota Land Cruiser. For the first time, regular patrols got under way and the arrests of suspects rose steeply on the ranch.

Over the next two years, Galana's elephant population numbered between 4,000 and 5,000, according to a joint aerial count I flew with David Sheldrick. Most of the tuskers had migrated from Tsavo East, we guessed, where the elephants had virtually eaten themselves out of vegetation. On Galana, they found plenty of water, abundant food, and a bit of protection. By mid-1976, they seemed reluctant to leave Galana because of the rampant poaching in the park.

In the past, the elephant herds had constantly been on the move between Galana and Tsavo, traveling long distances and allowing the vegetation and soil to recover. Now, however, they were seriously taxing the carrying capacity of the land. Irreparable damage was occurring to habitat, particularly to trees and shrubs that were important to our ranching, game, cattle, and safari operations. For the first time, we found ourselves trying to weigh how much game Galana's delicate soil could carry, especially in the arid north and west. We had already observed that in areas where cattle concentrated in tightly packed herds, grass cover quickly eroded, leaving a fine soil. The desiccating winds that blew five months a year could potentially carry off such soil, ruining Galana's delicate environment.

The habitat destruction caused by the large herds continued unimpeded, as did the killing of Tsavo's elephant population for illegal ivory. By the end of 1976, David Sheldrick, our friend and neighbor, was facing a flood of aggressive

Elephant herd gathered in the rainy season on Galana Ranch.

Somali poachers armed with AK-47s. He shockingly announced that over the previous two years, Tsavo had lost some 15,000 elephants. The disaster at Tsavo was compounded when the government, ostensibly to streamline operations, merged the national parks with the notoriously ineffectual and corrupt Game Department, whose scouts and top officials were involved in ivory smuggling. Dr. Perez Olinda, the first director of the national parks who had been trained as an ecologist, argued for independent overseers, but he was pushed out. Soon after, David submitted a confidential report on poaching and was also removed from office. He was named head of planning for the newly created Wildlife Conservation and Management Department. But six months after he was transferred from the park he had so long nurtured and protected, he died suddenly from a massive heart attack. David's forty-two-year-old widow, Daphne, was left alone to raise their twelve-year-old daughter, Angela, and to care for the orphan elephants she had been looking after for many years. At Tsavo National Park, David was replaced by a Game Department warden. The slaughter of Tsavo's elephants continued.

CHAPTER TEN

YEAR OF THE JACKHAMMER

IN 1977, ONE JARRING EVENT FOLLOWED ANOTHER. In early February our general manager received a letter from District Officer P. K. Gichuru in Malindi, headed, "Harrassment of Wananchi By Your Anti-Poaching Unit." Gichuru stated that his office had received complaints that the Galana Ranch anti-poaching unit was harassing local citizens and stealing their money. Our ranch manager, Ken Clark, supervised the unit and was very troubled by the allegations. He immediately asked the government to conduct a complete investigation and, upon discovery of sufficient evidence, to bring the culprits to criminal court. If the complaints proved frivolous, exaggerated, or fabricated, he wrote Gichuru, Galana expected public exoneration.

Ken was certain that he knew the origins of the complaints and that they didn't amount to much. In mid-January, our anti-poaching unit, accompanied by three uniformed rangers from the Game Department, had arrested four men armed with bows and poisoned arrows who were hunting illegally on Galana. During questioning by the game warden in Malindi, the suspects revealed places where trophies and illegal weapons were hidden. Because the government anti-poaching officers did not have a Game Department vehicle at their disposal, they asked the leader of our unit to drive them to the location in a ranch vehicle. The government rangers carried out three raids, ending at 3:30 A.M., in which they confiscated contraband and illegal weapons. They also arrested three men, all of whom were subsequently convicted. Our men never left our vehicle and did not take part in the raid. But it was no surprise that some of the locals tried to implicate our anti-poaching unit. In the eight years of its existence, our patrol arrested 135 poachers, of whom 107 were charged, convicted, and sentenced in court. During those same eight years, only two written complaints were filed against the unit, and both were proved to be frivolous. A little animosity toward us was hardly surprising, considering that approximately 90 percent of all those arrested were residents of villages adjoining our property. At least one male resident in almost all of the nearby villages had been arrested. Most of the poachers we arrested were dangerous, aggressive individuals who had attacked our men with knives, axes, and poisoned arrows.

In May 1977 we had another blow when the government imposed a total, permanent ban on hunting, including the capture of wildlife. Animal lovers and editorialists the world over cheered the ban, believing that at last something was being done in Kenya to halt the eradication of the last great game herds. We were shocked and discouraged by the decision, however. The ban forced 106 licensed professional hunters—including Mike Prettejohn and Ken Clark—to change their focus from rifles to cameras, hunt elsewhere, or fold their tents entirely. It meant the immediate end of our hunting safaris and a substantial loss of income, aggravated by our inability to shift operations quickly enough. A June safari of eleven people, including four who had planned to hunt, had to be canceled after the clients were already en route. Other hunting clients demanded the return of deposits that had been made on future safaris. The financial toll mounted despite our efforts to promote photographic safaris in place of hunting. We also feared that the poachers—without professional hunters, trackers, skinners, and clients in the field to report their activities to Kenya's understaffed and infamous Game Department—would be free to continue their wanton destruction of the country's wildlife. There was little chance that the ban would actually improve conditions for Kenya's game, given the extent of corruption and the probability that high-ranking officials and members of Kenyatta's family were involved in the illegal sale and exportation of skins and ivory.

As we feared, poaching increased dramatically. Soon, game manager Ken Clark was finding two, three, or four elephant carcasses a week, and an alarming number of armed Somali poachers were making incursions onto Galana. In late July, our poaching unit came across a band of twenty armed poachers in the north. A firefight broke out and the poaching unit called for reinforcements. By the time they arrived, led by Brian Heath, the shooting was over. Thankfully, no one was injured, but the risk to the lives of our men, as well as our game, was increasing rapidly.

A few days later, on August 3, 1977, a herdsman told Ken that a rhino had been killed near the Lali Hills area of Galana. With a pistol on his belt, Ken jumped on his motorbike with Bediva, a gun bearer, and took off to investigate. He was followed immediately by one of our hunting vehicles, which carried a tracker and a driver. Ken spotted the poachers as he approached Lali Hills at around 4:30 P.M. He ran over to the safari vehicle and grabbed a rifle.

As the poachers fled, Ken and the driver pursued them. In the ensuing firefight, one poacher was killed and two were wounded. Ken and the others then waited for our anti-poaching unit to arrive, not knowing if there were more poachers nearby and, if so, where they might be hiding. As darkness began to fall, they gave up waiting, and Ken got into the car with the Galana men. He stood up through the car's open roof hatch, scanning the terrain, while he decided whether they should circle the area to search for more poachers or go back to the ranch. Suddenly, two shots were fired at the driver's side of the vehicle. The driver swerved and saw another poacher forty feet ahead of them, taking aim. A split second later, the Somali fired. The bullet went through the windshield, struck Ken's gun belt as he was returning fire, then ricocheted into his chest and broke his back. The driver raced to the hospital in Malindi, with Bediva and the tracker cradling Ken in the backseat. The fifty-seven-year-old professional hunter survived for only a few hours.

Galana was shattered by Ken's death. Mike, Gilfrid, and I agreed to take care of his wife, Janet, and their two children, but we were well aware that nothing would ever ease their loss. Ken's well-attended funeral in Nairobi received a lot attention in the African and European press, and for at least a month afterward, Kenya's tourist trade slumped dramatically. And still the poaching continued. Seven rhino were found shot on Galana the week Ken died, and as many as fourteen elephants were slaughtered in a single day.

The Tsavo anti-poaching unit and a field force of the General Service Unit (GSU)—a paramilitary branch of the Kenyan Military and Kenyan Police— soon moved into the Lali Hills area and stayed there for several weeks. Although they did little to actually investigate Ken's murder, they beat up a lot of people, including a large number of our cattle herders. They also rounded up 120 people, shot 7 dead, and captured and wounded several others. The government used the shooting as a rationale for scrutinizing Galana Game and Ranching. Things were changing fast, and we had little understanding of the events that were enveloping Galana.

TWENTY-THREE DAYS AFTER KEN'S DEATH—on Friday, August 26, 1977— Illie and I were faced with another disaster. We were in the airport in Frankfurt, Germany, waiting to depart for Africa and anxious to get to Galana, when I received an emergency call from Heavenly Valley. Hugh Killebrew, my partner in the ski resort, had died with three other resort employees in a plane crash soon after they left Tahoe, bound for San Francisco. They were barely aloft when a second plane took off, and the two aircraft collided in midair.

Hugh's twenty-three-year-old son, Bill—a recent graduate of the University of California business school—was supposed to have been on that plane, but he was late. Hugh, who waited for no one, took off without him. Discovering that he'd missed the flight, Bill got in his car and started driving down to San Francisco. He soon hit a police roadblock, where debris from his father's plane had rained down onto the highway. Bill was one of the first civilians at the site of the wreck.

It was the second terrible event within a month. I had lost another friend, and Heavenly Valley had lost its managing partner and driving force. After two seasons with almost no snow, the resort was millions of dollars in debt. Hugh and I had been planning to raise capital to build an extensive snowmaking system to avoid a third catastrophic year. Bill and Hugh's widow, Eleanor, wanted me to come to Heavenly Valley right away to help sort things out. With winter rapidly approaching, they hoped that I could help oversee construction of the new snowmaking system. But I had no real choice. I had to continue on to Galana. Our game manager has just been killed in a gun battle, and Galana meant everything to me.

I phoned one of Heavenly Valley's remaining managers, Stan Hanson, who assured me that he could handle things until I got back from Africa. Then I called young Bill Killebrew and asked him if he could take the reins at Heavenly. He said that he could. Thanks to Bill's efforts, Heavenly Valley installed the snowmaking system and rebounded beyond anyone's wildest dreams.

Illie and I, meanwhile, continued to Galana. It was a sad time. Grief over Ken's death, combined with the ongoing aggression of well-armed and organized Somali poachers on the ranch, created a dark mood. It was intensified by government criticism over Ken's death. Officials were questioning why we had an armed anti-poaching patrol—dismissing the fact that the government had required an anti-poaching unit in our lease. As a result, we disbanded the patrol, despite the dangers of doing so.

By June 1978, the poaching situation was extremely serious for those who living on the ranch. Somali gangs carrying automatic weapons began roaming the property at will. Gilfrid was working very hard to get Kenya's paramilitary General Service Unit involved. He developed a close relationship with Ben Gethi, the head of the agency, and they planned a GSU operation that would take out about three hundred armed Somali poachers. A date was specified on which several hundred GSU police would travel overnight from Nairobi to Galana and begin an action at dawn.

The evening before the operation was to take place, however, Gethi telephoned Gilfrid to tell him that the action had to be called off and that Gilfrid would soon learn the reason why. The next day, on August 23, 1978, the radio announced that eighty-four-year-old President Jomo Kenyatta had died of natural causes. Kenyatta had been elected to a third presidential term, running without opposition, in 1974, but he had been enfeebled for years. After he suffered a series of strokes in his seventies, it was widely rumored that his wife, Mama Ngina, and his handlers had taken control of the government. Kenya had essentially become a one-party state, rife with patronage and tribal favoritism. Still, it was a relatively stable nation, and Vice President Daniel arap Moi succeeded Kenyatta in an orderly manner. We were grateful for the smooth political transition; we had other, more immediate crises to deal with at Galana.

After two years of welcome rainfall, the high grass on the ranch was fueling an outbreak of frightening fires. We set a few blazes deliberately in a strategy of controlled burning, and some occurred spontaneously. Others, however, were set by malevolent poachers. By July, Galana's staff was anxious and alarmed, and our managers were thoroughly exhausted.

Then, in August, poachers at the ranch went on an orgy of elephant killing. They left two hundred elephant carcasses—their tusks hacked out of their heads—rotting around Galana like outcroppings of bloody boulders.

We recruited retired game warden Bill Woodley to create a new anti-poaching unit, and we gathered all of our herdsmen for a very difficult, discouraging meeting. Almost all the herders confessed to having helped the Somali poachers. The news was devastating to hear—but what else, we wondered, could they have done, faced with so many men carrying so much weaponry? Woodley's subsequent interrogations led to the roundup and arrests of fifteen poachers, as well as the confiscation of eight automatic weapons and three hundred rounds of ammunition. This victory gave us some reason to hope that the situation might possibly improve.

Fortunately I have an optimist's eye. I told myself that the secret to life is to absorb the sorrow and move confidently toward better times. Still, I wrote in my diary, "Poor Galana."

CHAPTER ELEVEN

SLEEPING SICKNESS

EVERYTHING WAS CLEAR IN THE MORNING—the air, the birdcall, and my perception of problems—I thought as I was jogging along the road from Galana Lodge one day in March 1979. We had pulled back from the brink, and I felt as if the pendulum was finally swinging again in the right direction.

In the eight months since President Moi had taken office and our new anti-poaching unit had gone to work, the rampant killing of game was finally on the decline. I shared the same hope held by most Kenyans—that Moi would, as promised, root out corruption in government. A former schoolteacher, Moi had been Kenyatta's most trusted non-Kikuyu advisor and was widely respected for his integrity. He had assumed power without bloodshed, lifting the country's spirits. He had already publicly denounced five MPs for illegal practices and launched probes at the Works and Lands ministries. From the beginning of his presidency, Moi had reassured the public by going out into the countryside and speaking directly to the *wananchi*. He coined the slogan *nyayo*, meaning "footsteps" in Swahili. He eventually called it a philosophy of peace, love, and unity.

Illie and I had other good news. Christie and her husband, Rick, had brought our granddaughter, Heather, into the world. We immediately started recalibrating our lives and schedules. We sold our house in San Francisco, and I was planning to reduce my legal load. I wanted to spend more time, instead, on my own partnerships in Heavenly Valley, various real estate enterprises, Hawaii Airlines, trade publishing, Hawaii's first geothermal company, and, of course, Galana.

We had a capable new manager on the ranch, Jim Howard, and Brian Heath was doing an excellent job. Galana was better organized than it had ever been. We had greatly improved our water infrastructure, and the value of our fixed assets—buildings and improvements—had doubled. We soon diversified, too, by building a first-class lodge on the coast of the Indian Ocean near Malindi.

It was a beautiful location, with great marlin and other deep-sea fishing, and the four-acre property formed the northern boundary of the Marine National Reserve. The lodge we built was an exceptionally lovely Arab-style house on a promontory overlooking the ocean. Situated within a walled garden with bougainvillea, it had whitewashed walls and open arches that contrasted breathtakingly with the blues of the sea and sky and the pinks of the bougainvillea.

Our building contractor was Bunny Allen, a legendary white hunter in his early seventies who was known for his success in pursuing big game and beautiful women. Allen had come to Kenya from England in 1927. Within a year, he was the second gun to Denys Finch Hatton on a safari headed by the Prince of Wales. Allen later supervised one of the largest, most famous safaris on the continent—organizing a movable city of three hundred tents, twenty white hunters, countless bearers, and more than a thousand Samburu warriors—for John Ford's 1953 movie *Mogambo*, starring Clark Gable. Allen served as Gable's stand-in for dangerous shooting scenes, and he was famous for his reported romances with Grace Kelly and Ava Gardner.

I hoped that the lodge would be a place where our managers could get an occasional break from the stressful, arduous life at Galana. It could also house guests for photographic safaris, generate additional income from paying guests, and add another pleasant dimension to our trips to Africa. Going from the red clay dust and sparse vegetation of Galana to the clear light, cool breezes, and spacious rooms of the Malindi lodge was, for most visitors, restorative and refreshing. For me, it was a brief respite from Galana's challenges.

We now had 20,000 head of cattle on the ranch. Despite our progress, though, it was clear that ranching on Galana would never be a lucrative proposition. It was generating a meager profit of about one percent on our investment—considerably less than we could have earned on a certificate of deposit. In order to grow our enterprise, we were reinvesting all our profits in the ranch, and the founder-directors never took any compensation. There was, in truth, no real purpose to Galana except as a way of life and an effort to help the young Kenyan economy.

One reason there was so little profitability on the ranch was the expense of trying to prevent trypanosomiasis, the deadly sleeping sickness. The disease, transmitted by the tsetse fly, is carried by a parasite that lives in the fly's intestines. The parasite quietly slips out of the fly into a new host when the insect sticks its proboscis into a victim, human or animal.

According to the World Health Organization, 60 million people are currently at risk of sleeping sickness, and between 300,000 and 500,000 are infected at any given time. Without proper diagnosis and treatment, the disease is invariably fatal to human beings and causes the deaths of about 50,000 people a year. Unfortunately, its treatment and eradication have never been a high priority for policymakers or pharmaceutical companies, even though it ranks second only to malaria as a cause of death in Africa.

The agricultural effects of trypanosomiasis have been devastating to the continent's progress. It has been said that the disease is the main reason for Africa's failure to develop agriculturally. Its good ranch land is equivalent to an area twice the size of the United States. Much of it goes unused, however, because of the fly-borne parasite. Without the tsetse—which primarily inhabits the middle of the continent between the Sahara and the Kalahari deserts— Africa would likely be the meat producer for the world. Cattle would cover land that is now populated by elephant, lion, zebra, rhino, gazelle, and other wildlife.

Trypanosomiasis was a grave issue for our cattle. At first, when we grazed our herds, in dry seasons, within three miles of the Galana River, their trypanosomiasis infection rate shot up 30 percent and they started dying. We quickly changed the prophylactics we were experimenting with, doubled up on treatments, and learned to keep our cattle away from the river. Tsetse flies won't venture into open country, because they die quickly in the heat. Instead, they prefer cool, shady vegetation. Fueled by an amino acid called proline, the flies can reach speeds of up to fifteen miles per hour, but they can fly for only two minutes before they have to rest for an hour and refuel by making more proline. As a result, tsetse flies spend a lot of time hanging onto the undersides of leaves and patiently waiting for an animal, preferably a large, dark-colored one, to present itself.

In our early years, we received some excellent advice from an old veterinarian. We had been keeping trypanosomiasis under control in our herds by administering prophylactics and chemotherapy to infected animals. As we began experiments with preventive chemical dippings, however, the vet warned us against overtreating the cattle. Medications, he told us, would lower their natural immunities. Instead, he said, we would be wise to remember Charles Darwin. We should breed those animals that showed resistance—or trypanotolerance—and build our stock on their inherited DNA. We followed the old vet's advice and eventually cut the mortality rate of our cattle nearly in half.

In 1980 we also started an aggressive and successful research program with the Kenya Trypanosomiasis Research Institute (KETRI). To help KETRI unlock the secrets of tsetse control and the heritability of trypanotolerance, we built a modern research laboratory and permanent housing for KETRI scientists on the ranch. Initially, the KETRI team—led by Tom and Rosemary B. Dolan and Paul Sayer—monitored about nine hundred head of our Boran cattle, which we had dedicated to their research. Approximately a thousand years ago, the Boran, which originated in the highlands of Ethiopia, had migrated south and east with pastoralist tribes. One group of pastoralists settled in the tsetse-free Kenya Highlands. Others settled with their Boran in the tsetse-infested Tana River Delta of Kenya. Those cattle were known as the Tana-land Boran—or the Orma Boran, for the purposes of our study.

Our Galana-bred steers, which were originally from the Kenya Highlands, reached weights of 770 pounds, or 350 kilos, in three and a half years. The Orma-bred steers, which we purchased from the Orma tribe, took about six months longer to reach that weight. They also had some resistance to trypanosomiasis. Over the next several years, in areas of the ranch that had a low density of tsetse flies, we stopped all prophylactic treatment of the cattle. Our staff were trained to examine blood slides and take smears from sick animals to test them for tick-borne diseases and trypanosomiasis. We were able to eliminate wholesale monitoring and treatment of some herds until a test proved positive. That promised to cut costs and meant that we were doing less to compromise the cattle's natural defenses.

Unfortunately, in 1984 we encountered the first signs of resistant strains of trypanosomiasis. As a result, Galana had one of its highest infestation rates that year, even in places where tsetse had not previously been a challenge. When there was a high incidence of trypanosomiasis, animals all over the ranch tended to be in a weakened condition and susceptible to internal parasites, which contributed to the higher-than-normal death rate.

Then, on one night in 1987, 1,500 head of cattle died. All had been injected with prophylactic and curative drugs. The die-off was a tragic, bewildering event, and the financial loss amounted to more than $75,000. Although we never completely understood the cause, Brian Heath believed that the frequent and combined drug treatments—Samorin followed by Beneril—were an underlying factor. A similar incident had occurred in Zimbabwe twenty years earlier.

Brian and his wife, Sue, were working closely with the KETRI team and made a research trip to Zimbabwe. When they returned, we began experimenting with an intriguing method for preventing trypanosomiasis that had been developed two decades earlier by Dr. Glynn Vale, who had headed Zimbabwe's tsetse control efforts. Under the plan devised by the Heaths, we created a quarantined area—nearly 100 square miles of land that was densely inhabited by tsetse flies. We built roads crisscrossing it, brought in a water pipeline, and built staff housing. Brian and Sue then fabricated and installed 600 tsetse "targets." Each was constructed out of a square meter of black cotton cloth. An equal amount of black netting was attached to either side and stretched across a metal frame, which was then attached to a metal pole set in the ground. The frame moved with the breeze, attracting tsetse flies with its dark shape and the chemical scent of bottled phenols and ocytol. Research showed that the flies found black and royal blue irresistible. They would eventually fly into the netting, get a dose of insecticide, and then fly off to die. One of the traps was set up to collect the flies, and Sue Heath gathered them for study every week. The target-and-trap method showed spectacular results. During the first six months, not one of the two thousand cattle we kept in the quarantined zone became infected with trypanosomiasis.[1]

In 1985 we created a new quarantine area, complete with borehole, tanks, trough, pump, cattle yards, and an attendant's house. We also built two new airstrips, bringing the total to seventeen; an aircraft hangar; a two-million-gallon dam; two smaller dams; a new office addition; a digital weighbridge; and four houses for senior staff. Our asset value was growing at a robust 10 percent a year, and we were regaining our optimism. We took out some more big loans from the African Finance Corporation, which administered the funds for the World Bank, to fully capitalize and try to bring our herd up to our required maximum capacity.

Morale at Galana was at an all-time high. Personally and professionally, however, I was in a period of transition. As I turned sixty, I was fit and jogging, but I questioned the wisdom of running a marathon that year. I was thinking about mortality. "Life is finite," I wrote in my diary. "Plans for the next two decades are important." I wanted very much to do ranching the right way. "It is not enough to be the largest ranch producing beef in Africa. We must do it well," I wrote, "setting an example for labor relations and winning the government over to our economic philosophy of free markets, consistent as possible with the African tribal culture." But my overarching goal was to establish a way of life consistent with my love of open spaces and my desire to preserve them and conserve wildlife.

We were committed to protecting wildlife on Galana, including the lions. Many conservationists believed at the time that shooting any wildlife, including cattle-killing lions, should be forbidden. Others tried to promote the co-existence of ranchers and predators. We always tried to strike a balance, going

1. "Success in the Wilds of Galana," *Weekly Review*, September 15, 1989, 20–24.

after only rogue lions that killed cattle, and in the early 1980s we were losing some 300 head a year to lions.

In November 1982, the U.S. ambassador to Kenya, William C. Harrop, visited Galana for a few days. American ambassadors commonly visited Galana during their tenures to check up on us and make sure we weren't doing anything crazy out there on our one percent of Kenya. Harrop's visit was very pleasant. Brian gave the ambassador and his wife a few drives, day and night, to see the game and land under our stewardship, and they heard the characteristic coughing of lions around the camp at night.

Soon after, the herders reported that a lion had killed a prize bull twenty miles east of our headquarters. Most of the carcass was still intact, and the herders covered it with brush so that vultures would not get at it. Brian was normally in charge of killing lions on control, but he and Gilfrid were busy elsewhere this particular day. Mike Prettejohn was on Galana with his stepson, Harry, who was an avid photographer. The two of them, along with a couple of herders armed with spears, decided to sit up and wait for the lion by the carcass that night. Mike didn't have his rifle with him, so he borrowed Gilfrid's .450-caliber double-barreled gun.

I cannot improve upon Mike's description of his "lion on control" experience[2]:

> On arrival we found the carcass in the middle of a plain with little bush around. We could not build a "blind" too close for fear of being detected. The lion would remember if there had been any cover nearby. By dusk we had constructed a good blind . . . some distance away. We removed the brush off the animal and we sat ready with rifle and spotlight. Normally we would try to save the lion for a client if it was a good trophy animal, but on this occasion we had no clients. At about 8 pm two huge lion appeared at the kill. I decided to take the one with the least mane but nevertheless a very large beast. It was pitch dark by this time and the moon had not risen yet. As we heard the munching and tearing of flesh, I whispered to Harry to switch on the light. The two magnificent beasts raised their heads and turned as if to go. With only open sights it was quite difficult, but I took a shoulder shot and heard the bullet slap home.
>
> There was a mighty roar, the lion leaped five feet in the air and was

2. Michael Prettejohn, *Endless Horizons: 100 Years of the Prettejohn Family in Kenya* (Kenya: Old Africa, 2012).

gone. I believed he had a heart shot, but with cats especially, one tended to let them stiffen up before following up. We drank a cup of hot coffee before getting to the car to follow the spoor. We had no trouble in following and only a short distance away we could see the lion in the headlights. He was on his side but I could see his stomach heaving so I knew he was alive. This surprised me as I thought the shot better placed. I did not think he could move much. So I left the car with the headlights on him, crept out behind him with the intention of killing him with a shot in the back of the head. Before I had time to fire, he let out a terrific roar, did a backward somersault with his tail driving him like the propeller of a plane, and pounced on me! I fired once before he actually descended on top of me, but this shot only broke his foreleg and paw.

Like lightning the lion bowled me over. One paw almost ripped my left thumb off as I tried to hold him off my face and chest. I tried to shove him off with my left foot. I was now on my back while the lion took a huge bite into the calf area of my leg. The sensation was of having my leg tightened in a vice. This whole episode happened in seconds, and luckily for me Harry was taking pictures from the car with an automatic flash. The lion must have thought the flashes meant another shot was coming for it leaped away after the one bite (or was it the "taste?"). I gathered myself together and hobbled back to the car. I worried about septicaemia, gangrene and loss of blood.

We used the herders' *shukas* to try to stop the flow of blood and Harry drove for base as quickly as possible. It was quite difficult in the pitch dark to find the way across country to the nearest road. There were several different opinions as to the right way. As it transpired Harry chose the right way and we arrived back at HQ within an hour. Once there Gilfrid and Brian administered first aid. First aid during my grandfather's time would have been a feather dipped in iodine and drawn through the wounds. We had the use of cattle antibiotics and proper dressings. However, I was still losing blood, and could no longer bend my leg. So Gilfrid decided we should fly to Mombasa, about a 40-minute flight in the dark. The moon was now up, so we removed the back and middle seats of the Cess 206, put down a mattress and arrived in Mombasa just before midnight. During the short flight Gilfrid arranged by radio for an ambulance. On arrival I was whisked to hospital and by midnight I was on the surgeon's table with blood transfusion and all the antibiotics required. It took ten days before I was released and much of that time I was suffering from "kit-cat fever" in spite of all the antibiotics.

Mike Prettejohn under attack by a wounded lion during nighttime lion control.

The airport would not let Gilfrid fly back that night but at dawn he took off and returned to the scene to finish off and collect the lion. While in hospital an old settler visited me and said I was lucky to have modern medicine. He recalled his early days on the farm when his mate had been chewed by a lion. It took three days by ox wagon to get the man to hospital in Nairobi. There he had his leg tied to the ceiling for a month while all the muck was drained and in the end the doctor still had to amputate his leg. I certainly appreciated the access to modern medicine. This saved my life.

GILFRID RETURNED TO GALANA THE NEXT MORNING and killed the wounded lion. If we didn't deal with a cattle-killing lion, we could—and on several occasions did—lose herders to lion attacks. We almost lost Mike, and our herders were constantly vulnerable, living with the cattle twenty-four hours a day and sleeping on the ground in their small tents. But we did not kill many lions. In a year in which we lost three hundred head of cattle, we shot twenty-five lions, and the number we took on control had no effect on their overall population at Galana.

We were committed to wildlife protection, and in 1983 we drafted a proposal to designate most of the ranch—1.2 million acres—as a conservation area. We hoped to introduce controlled hunting, put radio collars on all the rhino at Galana to monitor against poaching, create a wilderness school where students would help run the conservation area, and hire conservation experts to oversee it. Poaching was becoming epidemic again, and I hoped Galana could be protected if we took the initiative.

The hunting ban in 1977 was, as we had feared, a recipe for disaster. It removed the watchful hunters from the bush—where it was in their self-interest to report poaching—and eliminated a way for Africans in rural Kenya to earn a legitimate living as professional hunters, game scouts, trackers, drivers, and hunting camp crew members. With those doors shut, poaching, for many, was the best opportunity for employment and a chance to cash in on the skyrocketing prices of ivory and skins.

Little could be done, I realized, in the face of the population explosion on the African continent, but we still hoped to make Galana safe for wildlife. I thought we would have thirty more years to try.

CHAPTER TWELVE

LOSING GROUND

ONE SUNDAY IN MID-OCTOBER 1985, we took a friend who was visiting from Hawaii on a scenic picnic out in the country. Henry Henley, who ran our photographic safari business, came along with us, and the outing began with lots of good animal sightings. The four of us spread out our picnic, enjoying the views and the sunshine. After about ten minutes, however, Henry said quietly, "Someone is watching us." He was looking over toward the bush. Three African men were gazing in our direction. They did not seem to be herders or local tribesmen. Even at that distance, there was something distinctly menacing about them.

Henry advised us to act normal. We began eating our lunch of cheese, carrots, and chicken, with little appetite. Then the three men—two of them carrying AK-47s—slowly started making their way toward us. Henry grabbed three bottles of Coke, and he and I walked forward to greet them. They were Somalis.

"*Jambo*," Henry said in Swahili.

"*Jambo*," one of the men replied.

As Henry continued speaking in Swahili, they fanned out in front of us, one to the left and one to the right. The man in the center was the only one who had some knowledge of that language. Henry walked toward them, holding out the Coke bottles. When he was about fifteen paces away from the group, the man in the center held out his hand, motioning that he should halt. Henry stopped, put the Cokes on the ground, and moved back a bit. The three Somalis walked forward and took the bottles. The leader then explained that they had seen Brian Heath shooting birds, and that they were doing the exactly the same thing—except they were shooting rhino and elephant. Government rangers at Tsavo and men in the anti-poaching units killed game, he said, so why shouldn't they? He said they were not interested in shooting us if we did not try to shoot them. Henry assured the men that we had no interest in hurting them. We smiled and said good-bye, and the Somalis left.

"I have never eaten so many carrot sticks," Illie later wrote in her diary, and they were hard for her to swallow as she watched the scene. It was an unsettling encounter, especially since Ken Clark's death, years earlier, was still fresh in our minds. And it signaled a drastic increase in poaching on the ranch. Because of the ceaseless poaching in Tsavo East National Park, the elephant population there was disappearing. In 1970, Tsavo East had had about 40,000 elephants, but barely 5,000 would be counted there twenty years later. About a third of the deaths were due to malnutrition and drought, but the rest were the result of poaching that had been unchecked since David Sheldrick died in 1977.

The devastation was not unique to Tsavo or Kenya. By the late 1980s, the poaching of African elephant was a continent-wide crisis. From Angola to Tanzania, armed ivory hunters were slaughtering elephants and bringing their populations to the verge of collapse. More than 100,000 elephants had roamed Uganda and the Sudan, but poachers reduced their numbers to about 10,000. In Tanzania's Selous Game Reserve, a decade of nonstop poaching had wiped out nearly half of its 100,000 tuskers. In Kenya, barely 20,000 elephants remained of the estimated 175,000 that had roamed the country in the late 1970s.

The ivory trade was merciless. Once poachers had killed all the mature bulls for their large, high-priced tusks, they began slaughtering cows, settling for whatever cash they could get for the lighter ivory. In 1979, poachers had killed 54 elephants to harvest a ton of ivory. Nine years later, they killed around 113 elephants for the same haul. The killing of cows decimated the herds. Calves under two years of age stood little chance of surviving if their mothers died. When researchers factored in trade and elephant statistics—age structures, reproduction, mortality rates, and tusk growth rates—the trajectory was undeniably clear. Africa's elephants were approaching extinction.

Little or nothing was being done. The Convention on International Trade in Endangered Species of Flora and Fauna (CITES) was impotent and toothless, and our urgent pleas to the government to help stop the poaching crisis on Galana met with official silence. Then, inexplicably, our relations with the Moi government began to fray. The first signs were its failure to pay the money it owed to Galana Game and Ranching as partners in the operation, with a seat on its board of directors. In January 1986, the government's Wildlife Conservation and Management Department purchased thirty oryx from our discontinued wildlife domestication program, but the government sent neither payment nor response to our monthly billings. When Brian Heath visited the district manager in Nairobi to ask for payment, he was assured that it would be forthcoming. Eight months later, we had still not received it. We had also been submitting annual compensation forms for cattle that were killed by lions, but we had never received payment for any of those billings.

These issues were not minor. We were operating a huge ranch at a one percent profit. All unnecessary losses were unacceptable, especially when they were caused by bureaucratic bungling or delays in a venture that had been launched with the government's encouragement and participation. We were endeavoring in every possible way to fulfill the conditions of our lease, but the government was increasingly uncooperative.

By 1986, poaching at Galana had reached intolerable levels, and the local police response was utterly ineffective. When one truckload of officers came upon a large gang of Somali poachers skinning a zebra on our northern boundary, they unleashed a barrage of gunfire without wounding or capturing a single poacher. Gilfrid, along with Mike and his brother-in-law Tony Dyer, had numerous meetings to discuss the crisis with an official at Hola. At each meeting, the official became more unpleasant, more uncooperative, and more critical of Galana. We didn't realize that he was actually helping the poachers, transporting their ivory in government vehicles and destroying Galana's reputation in Nairobi. He was apparently passing along bogus reports to cabinet ministers close to President Moi that our staff was secretly communicating with ivory dealers in South Africa and smuggling ivory out of the country on aircraft.

Soon, Brian Heath discovered a pile of twenty-five elephants—a whole herd—that had been slaughtered in one spot. Galana's remaining elephants fled in fear of the poachers, and when Illie and I arrived at the ranch in early October we saw none in the cattle ranching area. Gilfrid, Mike, and Brian had heard that the government was going to launch a secret offensive against the poachers, and we all hoped that would be the case.

Two days after our arrival, a twenty-five-man armed government security force arrived at the lodge. The captain, a General Service Unit officer, informed Henry Henley that about a dozen trucks loaded with men would be on the property for several days. We assumed that, thanks to the help of some influential Kenyans we trusted as well as other friends—including Prince Bernhard of the Netherlands, the former chairman of the World Wide Fund for Nature International—President Moi and his minister of Wildlife had learned about the slaughter of elephants on Galana and their need for protection. We guessed that the president had reacted positively and was taking action.

To our relief, the government soldiers who showed up on Galana—120 from the Kenya Army, GSU, the Criminal Investigation Department (CID), and provincial police—began a coordinated sweep from Tanzania to the Somali border on the coast of Kenya. There had been nothing like it in ten years. Their mission was daunting, however. The Somalis lived in the bush with no fixed camps and moved about in units of two to twelve men. The government soldiers, by contrast, were trained to move as a lumbering unit, with lorries transporting truckloads of men under the direction of a single officer. They badly needed to operate in smaller groups under unit leaders with sergeants. At their own risk, our herders and ranch managers began using ranch radios to give the government men up-to-the-minute reports of daily contacts with poachers. The Kenyan soldiers, however, failed to kill or wound any of the poachers, though one of their own men was killed and two were wounded.

Despite this pathetic display of skill and strategy, we felt slightly encouraged. Other events, however, were very troubling. Later on, a truckload of Kenyan army and police officers drove onto Galana, unannounced, and surrounded the home of Brian and Sue Heath. Several men, including a CID officer, entered the house and instructed Brian to open his gun safe. Inside, they found a nonfunctioning dart gun that John King had left on the ranch around 1970, when he was a veterinary officer and head of the Game Department's Capture Unit. The gun, which had not been used in years, delivered tranquilizing darts to humanely capture animals for veterinary treatment. The officers also found two smoke grenades that we kept in case of aircraft emergencies. Unfortunately, the grenades, classified as firearms, were in the house, not in the aircraft. The police confiscated them as illegal weapons. A few weeks later, Brian was officially charged in Malindi with a firearms violation for possessing the dart gun. Meanwhile, government men had also searched the house of our administrator, David Taylor. They discovered ham radio equipment. Since it was unlicensed, they said it was a violation of national security and charged Taylor with a crime, too. We were incredulous and hopeful that the absurd charges would be dismissed.

At the end of October, Brian and I had another run-in with Somali poachers. This time they were even more threatening and direct. He and I had flown to the northern part of the ranch, some forty miles from headquarters, to do a monthly cattle count at a *boma*. When our Cessna 206 came to a halt, the *boma*'s headman ran out to meet us at the end of the landing strip.

"*Bwana*," he said quickly, "there are bad men, Somalis, waiting for you over there! Go!"

I turned to Brian and said, "There's no point in trying to leave. They would have us before we got the plane turned halfway around. Let's go talk to them."

So Brian and I walked toward the Somalis and stopped a few feet away. One of the men explained very precisely that he and his men would kill Galana personnel if government soldiers did not retreat and leave them alone.

"Why do you put the Kenya soldiers against us?" he asked. "We take what God places here for all."

When Illie learned about this encounter, she was understandably upset. Still, we believed that with a well-planned, determined, and properly executed effort, poaching could be stopped without putting our lives at risk. When we returned to Hawaii in November, I decided to write to Prince Philip, the Duke of Edinburgh, and ask his assistance. I had met him six years earlier at a London meeting of the World Wide Fund for Nature International. Prince Philip, who was president of the organization, delivered a paper on the Atlantic salmon and described his efforts on behalf of elephants in Sri Lanka. I had presented a report on the elephant on private property in Kenya. Now I asked him to request that President Moi continue the anti-poaching sweep on Galana. I noted that in 1977, when our game manager was killed, there were 200 rhino and 5,000 elephant on the ranch. As of my letter, the rhino had been all but wiped out, and the elephant population now numbered less than 1,000.

"It is vital," I argued, "that the Kenya government keep up the pressure on the Somalis. To leave the field now might mean (1) the end of the elephant in Kenya's coastal area in less than five years except for those in a few parks and (2) the retaliation by Somalis against Galana's management and herders who have very actively assisted and directed activities by the Kenya forces against the Somalis."

We felt guardedly hopeful, but the scant optimism we had was shattered a few weeks later. In December, ministers of the Kenya cabinet ordered us to shut down Galana's photographic safari business. The order accompanied a letter accusing Galana Game and Ranching of posing a threat to Kenya's national security. There was no explanation for the allegation. As a result of the order, Jessica and Henry Henley had to cancel all bookings for Christmas 1987 and all of 1988, a major loss of income for the ranch.

That blow was quickly followed by another. Commissioners for the Tana River and Kilifi districts informed us that the government had withdrawn our permission to use Galana's airstrips—again citing national security as the rationale, with no explanation. We asked the government to reconsider. We explained to the provincial commissioners that without the use of the airstrips, we could not fly in food and veterinary drugs to staff when the rains began, supervise the construction of a school we were building, continue KETRI's fifteen-year trypanosomiasis research program, run our tourist business, or manage 24,000 head of cattle on the ranch.

To press our case, Gilfrid and Brian met with the provincial commissioner of Coast Province, S. P. Mung'ala, who repeated what was now a tiresome party line.

"As I told you," he wrote them afterward, "the Government is very much concerned with some of the businesses conducted by your Company within and outside the ranching area. Some of the activities are touching on State security and a threat to Kenya's precious wildlife." He informed us that the government was going to restrict our operations to cattle ranching and would not lift the restrictions on airstrips. We would be allowed to use one airstrip, but only after receiving explicit authorization and clearance from police headquarters in Nairobi for each use. Mung'ala's tone could not have been more hostile. He forwarded a copy of his letter to Joseph arap Leting, the permanent secretary to the Cabinet in the Office of the President, and to J. B. Kibati, the provincial security intelligence officer.

Gilfrid believed that the allegations were the result of the government's lack of understanding of Galana's history and all we had done. So he wrote to Paul Ngei, minister for livestock development and a political heavyweight. A former leader of the Mau Mau rebellion, Ngei had spent three decades in government as a member of Parliament and a cabinet minister for Moi and Kenyatta. Gilfrid wrote to Ngei that Mung'ala's December letter was "tantamount to an accusation against the Company of subversion, and willful activity, contrary to the interests of the State." No doubt his letter only ratcheted up the scrutiny of Galana.

Meanwhile, in a third effort to cripple our company, the Agricultural Finance Corporation foreclosed on a loan to us of some two million Kenya shillings and demanded payment in three days. With some difficulty, we complied.

The stakes only rose higher when Prince Philip, prompted by my earlier letter to him, wrote to President Moi, expressing his concern about poaching at Galana. If we were not already on the radar in President Moi's office, that letter may have elevated the ranch to an issue of direct interest to Moi himself and his closest ministers. Prince Philip's message to President Moi read:

Sandringham, Norfolk
5 January 1988

Your Excellency,

I think you may know that I am presently President of the World
Wide Fund for Nature (WWF)- International. In that capacity I have been
following the problems of poaching in East Africa with considerable alarm
and concern.

I thought you would be pleased to know that it has recently come to
my attention that the Government of Kenya has reacted strongly to the
poaching of elephants and rhino on the Galana Ranch by sending units
of the Army and Police to the area. I am told that this action is very much
appreciated by the management and staff of the Ranch. It has proved
to them that the Government is making a serious effort to prevent this
destruction of valuable wildlife, and this has had a significant effect on
raising their morale.

Their only anxiety is that this support might be withdrawn before the
poachers have been effectively driven off. They natu[r]ally fear that if they
are left to themselves, the poachers might well try to exact retribution for
the losses they have incurred.

I would just like to add that WWF is always ready to offer what help it
can in the conservation of Kenya's rich natural heritage.

Yours sincerely,

Philip (in signature)

I had badly miscalculated the political effect of Prince Philip's involvement.
I now believe it actually inflamed matters and possibly furthered an impression,
created by our detractors, that high-handed white foreigners—particularly an
American lawyer from Hawaii—were exploiting Kenya's natural resources.

We were fighting for Galana's life. Still, in our naïveté, it was difficult to let
go of the hope that common sense and goodwill would prevail and that the
value of our efforts to do the right thing would be appraised properly. Misguid-
edly, I enlisted the help of my old friend Charles B. Renfrew, former U.S. district
judge and deputy U. S. attorney general during the Carter administration, who
was now an officer and director of Chevron. Judge Renfrew wrote a four-page
letter to the U.S. ambassador to Kenya, Elinor Greer Constable. Noting that he

cc: HL tco

SANDRINGHAM, NORFOLK

5 January 1988

Your Excellency,

I think you may know that I am presently President of
the World Wide Fund for Nature (WWF) - International. In
that capacity I have been following the problems of
poaching in East Africa with considerable alarm and
concern.

I thought you would be pleased to know that it has
recently come to my attention that the Government of
Kenya has reacted strongly to the poaching of elephants
and rhino on the Galana Ranch by sending units of the
Army and Police to the area. I am told that this action
is very much appreciated by the management and staff of
the Ranch. It has proved to them that the Government is
making a serious effort to prevent this destruction of
valuable wildlife, and this has had a significant effect
on raising their morale.

Their only anxiety is that this support might be
withdrawn before the poachers have been effectively
driven off. They naturally fear that if they are left to
themselves, the poachers might well try to exact
retribution for the losses they have incurred.

I would just like to add that WWF is always ready to
offer what help it can in the conservation of Kenya's
rich natural heritage.

Yours sincerely

Philip

cc JE tco

Registered as
Fondo Mondiale per la Natura
Fondo Mundial para la Naturaleza

President
H R H The Duke of Edinburgh
Vice Presidents

and his wife, Barbara, had visited Galana and were impressed with the ranch's operation, he championed our efforts. "The history of Galana Ranch has been one of significant gains in game conservation and cattle production," he wrote. "It is a model for economic development that transcends cattle operations and game conservation. It illustrates how private resources and skills can work with Governmental officials and the people of developing countries in a cooperative venture to their mutual advantages."

Rudy Peterson—retired president of the Bank of America and former administrator of the United Nations Development Program (UNDP)—also wrote to Constable on Galana's behalf. He noted that he had stayed on the ranch numerous times and had, over the years, received UNDP staff assessments of Galana's activities. "They, likewise, were very much impressed with management's constructive handling of the Galana Ranch operation and in particular the experimental oryx domestication program," he informed the ambassador. "We all found the accomplishments of the Ranch remarkable and a model worthy of study for possible implementation in other areas of Africa in which the United Nations had a particular interest in developing protein production." As a member of the advisory panel for the East African Development Bank, he added, he would be willing to meet with members of President Moi's cabinet to discuss Galana.

Our misunderstanding of our difficulty with the Kenyan government was soon corrected. On February 25, Gilfrid went to Nairobi to meet with Harry Mutuma Kathurima, the personal assistant of Joseph arap Leting, secretary to the cabinet. After Gilfrid gave the assistant a full account of the issues that Galana faced, Kathurima appeared receptive. He advised Gilfrid to call the following week to schedule an appointment with Leting. After Gilfrid arranged the meeting, however, several government insiders sternly warned him not to meet with the secretary to the cabinet. They gave no reasons.

Disregarding their advice, Gilfrid confidently went to the appointment. Kathurima was the only other person present in the room when he met with Leting. Believing that the secretary to the cabinet had been fully briefed, Gilfrid casually offered to open the discussion. Leting abruptly cut him off. Referring to the government's 10 percent stock ownership of Galana, he said, "Why have you never paid any dividends to the Ministry of Livestock?"

Gilfrid explained that we had returned every cent to developing the area, and none of the shareholders had been paid any dividends.

"I cannot understand," Leting said, "that after twenty years you have never given the government any dividends. You only have the government shareholding to protect you. Why should the government be involved?"

"To ensure the Ministry of Livestock's active participation in the largest ranch in eastern Africa," Gilfrid answered.

Leting went on to ask when Galana started, why the government was not more involved in its management, and other rudimentary questions. He continued with a litany of wild accusations. "Why did you shoot one hundred lions in one months? My staff have calculated you could have killed twenty-two thousand lions."

"Mr. Leting," Gilfrid said, trying to correct the absurdity, "one hundred lions were shot over a period of seventeen years. We lose an average of twenty-three head of cattle to lions every month."

"Yes, I am told you push you cattle to where the lions are, so that it is easy to shoot them. You lure lions with your cattle," Leting charged. "You lure elephant by providing good watering points, so that Somali poachers find it easy to kill them. Why did you not prevent the slaughter of all the elephant? Twenty-six on Christmas eve, including baby elephant."

"Mr. Leting," Gilfrid objected, "we have been informing government since 1976 of the urgency and scale of Somali poaching." But Leting continued with a barrage of bizarre allegations, leaving Gilfrid at a loss for words:

- "Why were you so cooperative during the GSU operation? We could not understand why you were so helpful."

- "We have closed your airstrips. You could easily be flying out rhino horn and ivory."

- "Some members of the Cabinet want to settle Galana. Some people wish it to be a National Park, so that we can properly protect the game. We are undecided at the moment. What would you do?"

- "Galana is no use to Government. It would be different if you sold thousands of head of cattle overseas, bringing in thousands of dollars . . . .If you provided hundreds of thousands of dollars in foreign exchange, we would not mind."

Leting then began questioning Galana's three-decades-old articles of association. "These articles must be changed," he insisted, and he instructed Gilfrid to write a "white paper" for him on the future that we envisioned for the ranch.

"Of course, if we take it over," Leting added, "you will be properly compensated."

"Mr. Leting," Gilfrid tried to explain, "the main reason for seeing you is to try to prevent the adverse international publicity Kenya is receiving. As a fellow Kenyan, at all costs, we cannot afford to have law suits filed against Galana for breach of contract in our safari business. I am also concerned that, as Galana represents a large American investment, the American Ambassador may become involved . . . ."

"I do not mind about foreign opinions, where our precious wildlife is at stake," Leting interrupted. "And if you think you can continue to operate Galana because of American pressure, you will be operating under very unsatisfactory conditions. You write me a paper on how you think the ranch should operate," Leting stated, ending the meeting.

After that encounter, we were sure that the efforts to shut Galana down originated at the government's highest levels.

CHAPTER THIRTEEN

TOO BIG TO SUCCEED

SUMMONING SOME LAST VESTIGES OF HOPE for Galana's future, Gilfrid, Mike, and I put together a "white paper" for Leting and the other ministers on the history of Galana Game and Ranching. Perhaps there was still time, we thought, to overcome the ignorance of members of the government about the genesis, purpose, and evolution of the ranch. They were flagrantly unaware that we only rented the 1.6 million acres of the ranch and that the government would take it all back in 2012. The lies and fabrications about Galana were proliferating, and they were part of a campaign to get rid of the whites, or *mezungas*, who had been running it.

The white paper gave the original history of Galana's formerly unpeopled land, the government's goals, and the requirements of the lease, as well as evidence that we had surpassed each one of its stipulations. It chronicled the steady improvement and size of our cattle ranch. With a balanced herd of 26,000 head, it was the largest cattle ranch in Kenya and perhaps all of Africa. The white paper tallied the miles of roads, airstrips, buildings, boreholes, pipelines, and dams we had developed and the hundreds of thousands of dollars we had invested in their construction. It also provided annual accountings of game animals killed—before the hunting ban and afterward, on control—and details of our two decades of scientific research on the domestication of wildlife and the prevention of trypanosomiasis. It documented the successful transition of our hunting safari company to a photographic operation after the hunting ban went into effect in 1977. It demonstrated our employment of 400 people from a variety of tribes, and our Kenyanization of Galana's management. Finally, it showed that licensed safari hunting on Galana, when it had existed, had killed less than 5 percent of the 300 elephants a year that the government had wanted us to crop for meat production. It was a singular record.

We distributed the hastily prepared paper to various ministers and members of the cabinet, but it failed to stem the rumors about Galana.

IN MARCH 1988, after months of delays, Brian Heath finally got his day in court on a charge of possessing illegal firearms, stemming from the search and seizure of John King's tranquilizer dart gun. Illie and I flew in from Hawaii to support him. Since the case appeared to be motivated more by inscrutable political motives than by legal infractions, we hired Kenya's top defense lawyer, Byron Georgiadis, to defend Brian. After a couple of hearings, the investigating officer took Brian aside and confided that there were "instructions from above," from the president's office, that Brian was going to be found guilty, no matter what. He advised Brian to plead guilty or suffer a harsher sentence. After discussing the conversation with Georgiadis, Brian reluctantly agreed to change his plea. In return, he was assured that he would not be sent to jail. In the end, Brian was fined 20,000 shillings (about $250). David Taylor was also fined, even though Georgiadis proved in court that he did not need a license for his ham radio.

As it turned out, the search the government conducted had been part of a larger investigation of Galana, based on a torrent of misinformation—that we were involved in poaching, killing hundreds of lions, flying ivory out of the country, and other crimes. Another allegation was that Galana's herdsmen were aiding the Somali bandits, feeding them and allowing them to stay in their *bomas*. There may have been some truth to that, but it was hard to blame them. When Somalis carrying AK-47s marched into a *boma*, asking for *chai*, the herdsmen would have been foolish to provoke them. Moreover, their tradition of African hospitality would have inclined them to act generously to strangers.

Brian developed a reasonable relationship with the investigating officer. In the end, he showed Brian two very thick files, the result of a probe lasting more than a year. The officer said that he could find nothing to substantiate the allegations and had recommended closing the investigation. But it was too late. Leting had already told Moi that the allegations were true, and he was not going to back down.

In late March 1988, Moi was reelected president in a one-party election. In August, we received a three-sentence letter from the Kenyan commissioner of lands. It read:

Please note that the Government has decided to evoke [*sic*] Special
Condition No. 21 of the above mentioned lease. I am accordingly giving the

company the six months notice with effect from 11th August, 1988. To effect
this, the necessary steps will be taken in accordance with the provisions of
the said Special Condition.

Yours faithfully,

J.R Njega

Commissioner of Lands

Condition No. 21 of our lease specified that Galana Game and Ranching
Ltd. could be required to surrender the land for the construction of roads,
railways, school, hospitals, etc. on six months' notice. It also specified that the
company would be paid a fair and reasonable compensation for any permanent
improvements it had made on the property. If unchallenged, the six-month
notice to vacate meant we would have until February 11, 1989, to leave Galana.

In response to the notice, we threatened to bring a lawsuit against the
Kenya government. Gilfrid, worried that his interactions with officials had
provoked the government's action, resigned from the ranch's board of directors. To gather support for our position, Mike, Brian, and I met in October with Minister for Livestock Development Maina Wanjigi and U.S.
Ambassador Elinor Constable. The ambassador was very helpful. She was
concerned, she told us, about the process of law and the principles involved
in condemnation. If the move against Galana violated them, it might un-
settle foreign investments in Kenya. Minister Wanjigi, an assistant minister
under Bruce Mackenzie in 1967, was also sympathetic. He was well informed
about Galana's beginnings and said that there was no evidence that Galana
had been involved in poaching. It was, however, the government's position,
he explained, that the ranch posed a national security risk because of its prox-
imity to the Somali border. Still, he asked us not to file the lawsuit. I agreed
to stop legal action if the government would withdraw its February 11, 1989,
deadline for condemnation. Wanjigi agreed. Although Ambassador Constable
offered to arrange a meeting for me with President Moi, none was planned.
The ambassador later guessed that Moi had not made up his mind and was
avoiding a meeting—or perhaps he was not willing to change the order of
eminent domain, regardless of facts.

At the beginning of 1989, we were slightly hopeful when Mike Prettejohn
was able to secure a meeting with the Kilifi district commissioner. Mike went
to the meeting with Richard O'Brien Wilson, a well-respected rancher who was
chairman of the Central Agricultural Board. Wilson was also vice chairman
of the Agricultural Development Corporation, the entity responsible for the

Kenyanization of private lands after 1967. Mike tried to steer the meeting toward a forward-looking agenda. Government investigations, he said, had found no connection between ranch personnel and the Somali poachers. He argued that the government should reinstate our airstrip permits and cooperate with Galana on its future development. Mike's miscalculation, however, was evident as soon as the district commissioner began to speak. "This may be so now," the official stated, referring to the findings that absolved Galana, "but you have in the past been involved with illegal activities, and you cannot get away from the fact that you have had convictions in court in the matter."

Mike attempted to explain that while Brian was technically guilty, he had not really committed any crime. The district commissioner interrupted him:

"That is insignificant. The fact is that your people were proven guilty."

He then asked Mike several rudimentary questions about the ranch, such as its size and its number of cattle. When Mike attempted to answer them, the district commissioner abruptly interrupted him.

"How can your small company possibly operate and develop to the full such a piece of land? I consider it very underdeveloped."

The conversation was disappointing, but at least there had been some dialogue. Mike and Wilson left the meeting hoping that, at the minimum, they had opened a line of communication with a government official.

The date for the government's takeover of Galana—February 11, 1989—came and went without incident. Conditions on the ranch, however, were deteriorating. One evening, David Taylor was having a beer when a band of armed poachers appeared at his house. They marched him, in the dark, a mile and a half back to the office. There, they ordered David to open the safe, took two thousand dollars in cash, and marched him home. Some of the poachers ransacked David's possessions before their leader made them return the items. The Somalis also gave the herders a hard time, fired on Brian's plane, and generally terrorized everyone on the ranch. Most of the herders were now saying that they wanted to leave Galana, but Brian managed to talk them into staying. By the end of the month, he was exhausted, at the end of his rope.

On March 13, Brian, Henry Henley, and I got away for an afternoon of wonderful fishing off Malindi. When we returned to the Indian Ocean lodge, however, we got more bad news. Eleven more elephants had been shot near one of our water tanks. In just two days, poachers had killed a total of twenty-four tuskers. Mike Prettejohn spotted the carcasses from his plane and circled over them while a GSU anti-poaching squad went to the scene. When the GSU arrived, Somalis in camouflage suits beckoned them to come forward. The corporal who led the government force thought the Somalis were game rangers and walked forward with his six men. The poachers promptly shot the corporal dead and wounded another soldier in his right elbow; he would face the amputation of his arm. In all, the Somalis fired three shots, and the GSU fired one. None of the Somali shots missed, and the soldiers quickly panicked and fled. When reinforcements returned, it took them two days to find the corporal's body. They also recovered all of the poached ivory. The largest tusk weighed fifteen pounds, and the smallest tusk was so tiny that it was virtually worthless.

That same day, when Mike arrived at our airport office in Nairobi, he found an official letter from the commissioner of lands giving us six months' notice of condemnation, effective September 12, 1989. The government was condemning our roads and schools and our disease control, game preservation, ranching, tourism, and security operations.

"So, this is how Galana ends," Illie wrote in her journal. "If it is like ADC [the cooperative farm the government created from tribal lands] across the river, it will go to pieces." We were all in shock—Brian and I, perhaps, more than anyone. We had remained hopeful that we could work something out with the government. But Illie had been more skeptical. She was anxious about the dangers confronting us, and after my confrontations with the Somalis she was rightly fearful about what might happen. When Gilfrid resigned, she had guessed the end was near. Feeling sympathy for my stunned disappointment, she voiced her anger in her diary.

The government's position, we later learned, was more complex and destructive than we had imagined. While the GSU soldiers were putting their lives at risk, the government was actually involved at every level of the trade in illegal ivory. Richard Leakey, the renowned paleoanthropologist, revealed in his book *The Wildlife Wars* how entrenched poaching was in the Kenya Wildlife Department when he took charge of it in 1989. Poaching was rampant even among rangers who were supposed to be protecting Kenya's elephants and rhinos. "In numerous cases," Leakey noted,

> the tracks of government-owned Wildlife Department Land Rovers were found next to the carcasses of elephants. Casings from .303 bullets were even recovered near the dead animals. The rangers referred to these types of killings as "roadsiders," because the elephants had obviously been killed by someone pulling up beside them in a car and shooting them point-blank. It's difficult to say how many thousands of poached elephants were slaughtered in this manner, but I'm afraid it was a sizable proportion.

In March 1988, a two-day aerial reconnaissance in and around Tsavo East counted twenty-five fresh elephant carcasses. "Nearly all the animals," Leakey disclosed, "had been shot close to the road." One of his officers explained that it was motorized poaching by Wildlife Department personnel. "They were using government bullets, fired from government firearms, from government vehicles, driven by government drivers, and fueled by taxpayer's fuel. And the whole thing was being supervised by government officers, some of them being top wardens in the Wildlife Department."[1]

Smith Hempstone, the blunt and voluble U.S. ambassador and former editor of the *Washington Times*, arrived in Kenya in late 1989 and analyzed the poaching problem in his book, *Rogue Ambassador*[2]:

> Paradoxically, the situation worsened in Kenya after legal hunting was banned in 1977. The presence of trophy hunters, who destroyed their snares and provided the game department with intelligence, had acted as a deterrent to poachers. And the loss of fees from legal hunters deprived the game department of an important source of revenue. Poaching might still have been contained had members of the Kenyatta family not been deeply involved in the racket. As it was, the poachers killed with virtual impunity. When a significant number of poorly paid game scouts stopped protecting the animals and joined the killing, wildlife lost its last shield.

The rampant poaching became even more dangerous when poachers began killing foreign visitors, crippling Kenya's $350-million-a-year tourist industry. The first victim was a twenty-eight-year-old English wildlife photographer named Julie Ward. Her burned, dismembered body was found in September 1988 in Kenya's most popular tourist destination, the Masai Mara Game Reserve, a week after she was last seen there. The Kenyan police probe into the crime became an international embarrassment to the tourism-dependent nation. Authorities initially theorized that she had been eaten by lions and struck by lightning. Only after the victim's father, hotelier John Ward, uncovered further evidence did Kenyan authorities acknowledge that she had been murdered.

1. Richard Leakey and Virginia Morell, *Wildlife Wars: My Battle to Save Kenya's Elephants* (London: Macmillan, 2001), 64.

2. In addition to my memories of visits with the late Ambassador Smith Hempstone and our fishing trips on the Indian Ocean, I was grateful to have the benefit of his memoir about his time in Kenya, *Rogue Ambassador* (Sewanee, TN: Sewanee: University of the South Press, 1997).

Somali poachers were prime suspects in the subsequent killings of four other Europeans and the 1989 murder of a forty-nine-year-old American, Marie Esther Ferraro, of Bethany, Connecticut. Ferraro was killed and another American wounded when they were traveling in a Connecticut Audubon Society minibus near Tsavo West National Park.

A political, criminal, and emotional atmosphere was contaminating Kenya. We weren't able to see the full context of the motives and machinations that led to the end of what we had worked so hard, passionately, and hopefully to create. And little did we know how soon or how uncivilly it would arrive.

In March 1989, the district police officer in Malindi asked us to call a meeting of our Orma herders to explain that, for security reasons, the Orma people were going to be moved off the Galana Ranch. We assumed that the government meant this to be an informational meeting about an action planned for the future. Certainly our employees would have time to collect their belongings and say their good-byes; we would have time to find replacements to care for our 25,500 head of cattle and 300 camels; and we also assumed that we would be able to appeal the decision to the provincial commissioner in Mombasa or to Nairobi authorities. But as soon as the meeting got under way, without warning, the police began herding seventy men, at gunpoint, into a waiting truck. They were given no chance to get their possessions, identification papers, or anything else. Gilfrid was furious, and the police cautioned him against acting aggressively.

I climbed into the plane, flew to Malindi, and immediately called Mung'ala, the provincial commissioner in Mombasa. It was 4:15 P.M. when I reached him. Mung'ala listened to me, sounded sympathetic, and promised to telephone his commanding officer. I naively believed that officials would halt the kidnapping of the Orma herders. But the next day a truck arrived and took seventeen more Orma off the ranch.

Our situation was desperate. All but one of our camel herders had been taken away, and we had barely enough people left to care for the cattle. The district official in Malindi and his commanding officer did not like what was happening on Galana and apologized. But it was obvious that their hands were tied. The police were likely carrying out orders that had been orchestrated in Leting's office.

On Sunday, March 20, Mike, Gilfrid, and I flew to Nairobi for a 9:00 A.M. meeting with Kenya's police commissioner, Philip Mule Kilonzo. He listened attentively for an hour as we told him about the events at Galana. He was the finest Kenyan official I had met—professional, intelligent, politically astute, and well aware of citizens' rights. He was appalled at the way the police had treated our herders. Kilonzo called Mombasa's chief of police and told him that if he had to take the Orma off the ranch for security reasons, he should do so only after Galana received adequate notice, and he should treat the Orma herdsmen with courtesy. The commanding officer in Malindi agreed to delay any further action until Tuesday. He kept his word. Then, promptly on Tuesday, a government truck arrived, and the police loaded all the northern people, including the Orma, into the vehicle and took them off the ranch.

That night, we had dinner with many of our Galana friends and partners. "We all knew it might be our last dinner at the Lodge," Illie wrote in her diary. "We just can't believe it." She copied out a Kenyan saying: "Mountains can't be moved, but people can."

We had no choice now but to concede that the Galana experiment was over. Still, we tried until the end to do whatever we could to save the ranch, even as the days and weeks of our final six-month notice ticked away. We called on all of our closest associates to intervene. Initially, they would agree to help, but invariably, after looking into it, they would come back to us and say, "If it's about Galana, there's no way." In August, an Indian lawyer approached us and offered to guarantee an interview with President Moi and a reversal of the quit notice. His price was ten million shillings. We declined.

In those waning days, however, Philip Leakey—Richard's brother and an assistant minister at the Environment and Natural Resources Department— offered to arrange a meeting for Brian Heath with President Moi. Leakey had been reluctant to do so earlier, worried that he would be viewed as a champion for Europeans. Now, though, Leakey asked Moi if he would see Brian, and the president agreed. Brian received word one evening that he should be in Mombasa the next morning to meet with the president at the State House. Philip Leakey and Walter Kilele, boss of the Agricultural Development Corporation, were in the president's office when Brian entered. President Moi was dressed, as usual, in a dark suit.

The meeting was not long. Brian made a short statement about Galana and its situation. Leakey offered a few words. Then President Moi spoke.

"I was never informed that you are Kenyans. Why did you not come to see me earlier?" he asked Brian. He agreed to reconsider but noted that things might already have progressed too far.

We convened our board, and all agreed that the time had come for us to leave Galana.

It was too big to succeed.

EPILOGUE

TWENTY-THREE YEARS HAVE PASSED since the government of Kenya took over the Galana Ranch. Mike, Gilfrid, and I sold off most of our cattle and moved the remaining herd to a smaller ranch nearby. Hundreds of herdsmen returned to their homelands or went in search of work elsewhere. Our scientific research stopped. The astonishing adventure and outsized prospect that drew two intrepid Kenyans and one American—a hunter, a rancher, and a lawyer—to plunge into raw Africa came to an end.

I have wrestled long and hard with questions about our Galana experiment—what went wrong and why and how we might have avoided the pitfalls that eventually swallowed us. I believe that Galana, politically, was simply too big to survive. Kenya's population expanded geometrically, from 8 million in 1960 to 25 million in 1989. The fact that a few white men controlled so much land inevitably provoked distrust, jealousy, and greed. Combined with corruption, our own naïveté, and the conflict between hunters and non-hunters, it was a dynamic of destruction that condemned Galana and brought its elephant herds to the edge of extinction.

What we tried to do on Galana may never be repeated in Kenya. And I profoundly hope that Africa will never again see the decimation of its elephants that we witnessed in the late twentieth century. It is the duty of governments, conservation organizations, and proponents of hunting to come together and find a common path forward that will protect Africa's wildlife for the future.

Sadly, the often emotional rhetoric between the hunting and the non-hunting communities obscures a large area of shared agreement. Both camps want passionately to preserve animal populations. Both sides know that wildlife needs protection from a shrinking wildlife habitat and a growing human population.

But what type of protection?

No hunting, full stop? Or unregulated hunting, left to the individual's judgment?

Neither, of course.

A number of years ago, I had a discussion with anthropologist and wild-life conservationist Richard Leakey when he was in Northern California. The son of noted anthropologists Louis and Mary Leakey, Richard Leakey discovered the remains of "Turkana Boy"—a 1.6-million-year-old *Homo erectus* skeleton that was recovered virtually intact—and the 2.5-million-year-old "Black Skull," which forced reconsideration of the structure of the human family tree. Leakey's reputation and international stature grew when he published (with Roger Lewin) two best-selling books, *Origins* and *People of the Lake*, in the 1970s. During the 1980s, Leakey was the director of the Kenya National Museums and chairman of the board of the East African Wildlife Society, a conservation group. He was outraged by the slaughter of elephants in 1988, and the following year, he was appointed director of Kenya's Department of Wildlife and Conservation Management. Within a month of taking office, he concluded that all the rumors he had heard about the Wildlife Department were true. "It was," he wrote, "one of the most—if not *the* most—corrupt organization in the government."

Shocked, too, by the out-of-control violence against elephants and people—punctuated by the murder of revered conservationist George Adamson in August 1989—Leakey fiercely supported Kenya's ban on hunting and was a driving force behind a total international ban on the sale of ivory. To underscore Kenya's commitment, he held a bonfire, presided over by President Moi, in which they immolated the nation's store of elephant tusks.

When I met with Leakey, he insisted that the hunting ban was necessary to control the rampant poaching and slaughter of Kenya's wildlife. By late 2006, however, he had had a change of heart, as illustrated by his remarks at a wildlife conference in Nairobi:

> When my friends say they are still very concerned that hunting will be reintroduced in Kenya, let me put to you—hunting has never been stopped in Kenya, and there is more hunting in Kenya today than at any time since Independence. Hundreds of thousands of animals are being killed annually with no control. Snaring, poisoning, and shooting are common things. So when we fear the debate about hunting, please, do not think there is no hunting. Think of a policy that will regulate it, so that we can make it sustainable.

No one is more supportive of wildlife welfare than Richard Leakey. Surely, the time is long past for a reasoned, respectful consideration of wildlife hunting regulation, in consultation with expert wildlife biologists. I believe that we must do the best we can in setting licenses for hunting quotas, a position that Leakey later came to endorse. When local populations share in license revenues and acquire a vested interest in legal hunting, poaching is much easier to control.

It is important to note that with Leakey at the helm of the Kenya Wildlife Service—empowered by President Moi to root out corruption and to use force to stop poaching and the ivory trade—Kenya's elephant population has made a partial recovery. By 2008, the number of elephant in the Tsavo catchment had doubled, to 11,696, as tallied in the three-year census. Small elephant herds were seen again in the Lali Hills area. Still, it is a very fragile recovery. The mere rumor that ivory sales would be allowed again led to a resurgence in poaching.

There is talk that the government plans to jump-start a series of small ranches on Galana and resurrect the area's economic potential. I hope that is true. I write this on July 30, 2012—the very day that our original forty-five-year lease on Galana would have come to an end.[1] For me, the time we spent at Galana is now a lifetime away. It was an adventure, a way of life that can never be duplicated or repeated—just gratefully remembered. At Galana, we did our part for hunting and conservation, for science and land management, for domesticating animals, protecting wildlife, and preserving nature's magnificent landscape.

We worked a bit of Africa.

Safari masuri sana.

Kwa Heri

Martin Anderson
July 2012

---

1. Had the Kenya government, in 1989, fulfilled President Kenyatta's promise of *harambee*, it would now be taking over a successful Galana Ranch. What the Kenya government shut down was a viable operation that, by 2012, would certainly have exceeded the 26,000 head of Boran cattle, with a state-of-the-art veterinary and tsetse fly study team and a game domestication program that had significant recognition in the global wildlife community. No doubt the scientific work by veterinary doctoral candidates would have continued to contribute to the many research efforts in cattle diseases, tsetse fly control, and protein production.

LETTER OF ALLOTMENT

Tel: 87471: Ext.202

DEPARTMENT OF LANDS,
P.O. Box 30089
NAIROBI

Ref: 65678/IV/72

7th June, 1967.

Messrs:
MARTIN ANDERSON, ESQ.,
MICHAEL GWYNN PRETTEJOHN, ESQ.,
JOHN GILFRED LLEWELLYN POWYS, ESQ.,
c/o Messrs. Murdoch McCrae & Smith,
P.O. Box 65678, 6878
NAIROBI

LETTER OF ALLOTMENT

GALANA GAME/RANGE MANAGEMENT SCHEME

Gentlemen,

I have the honour to inform you that Government is prepared
to offer you a Grant in respect of the above-mentioned Scheme, of
an area of approximately 1,500,000 acres, on the following terms
and conditions, and subject to the charges specified hereunder:-

Area: 1,500,000 acres approximately as illustrated
edged red on L.D. Plan No. 65678/IV/47. The
Government will accept no responsibility for any
variation in the area which may be found upon
final survey.

Tenure: Leasehold under the provisions of the Government
Land Act (CAP. 280).

Term: 45 years, from 1st July, 1967.

Annual Rent: For the first five years of the term: K.Shs. 6,000/=
per annum. Thereafter, for the period from the
sixth to the tenth years of the term, inclusive:-

1. For Game and Tourist Development Areas:
(excluding hotel sites): 5 per cent of the
unimproved value of the land as assessed in
the fifth year of the term.

2. For Ranching Development Areas:
$2\frac{1}{2}$ per cent of the unimproved value of the
land as assessed in the fifth year of the
term, with allowance for deferment.

Provided that as between the sixth and tenth
years of the term inclusive, the total annual
rent for the Development Areas under 1 and 2
above shall not exceed K.Shs. 80,000/=.

Thereafter, for the residue of the term and for
the entire area:- 5 per cent of the unimproved
value of the land as assessed at the end of the
10th, 20th and 30th years of the term.

......./2

[...]

- 3 -

    (c)   establish a minimum of four permanent watering points or expend a minimum of K.Shs. 240,000/= out of the total of K.Shs.500,000/= on water development.

    (d)   construct a meat factory and refrigeration plant with such facilities and on such a scale as to permit of the processing of all game meat which becomes available as a result of game cropping and hunting operations within the agreed quotas in accordance with Special Condition No.16.

    (e)   construct administrative buildings; a dispensary; a school; hunters' lodges and a causeway across the Galana River at a site between the Lali Hills and Sala Hill, to be agreed with the Government of Kenya.

    (f)   construct such roads as may be necessary to provide sufficient access for the effective development of the entire area of the Scheme.

4.   The Grantee shall, within three months of the commencement of the term of the lease, appoint a manager who is resident within the area of the Scheme.

Special Conditions

1.   The Grantee shall, within nine months of the commencement of the term of the lease, promote a company with a total issued capital, either by loan or by equity, of not less than K.Shs. 2,000,000/= and the Kenya Government, or its approved agency, shall be allowed to subscribe to the issued capital of the aforementioned company in such sum as it may decide, but not exceeding 24 per cent of such issued capital. In the event of the Company increasing its capital beyond K.Shs. 2,000,000/=, the Kenya Government shall have the right of subscribing for up to 24% of any new shares it may wish to take up in the increased capital.

2.   The company so formed shall invest the full amount of its capital of K.Shs. 2,000,000/= in the development and administration of the land.

3.   The constitution of the Company referred to in Special Condition No.2 shall provide for the right of the President of Kenya at all times to appoint one member to the Board of Directors of the Company to represent the interests of the Government, and for the right of any organisation, which, on behalf of the Government of Kenya, subscribes more than 10% of the issued capital of the Company, to be similarly represented.

4.   The Grantee shall between the 5th and 20th years of the term of the lease carry out the following development, over and above the development of the area of 60,000 acres which is to be carried out during the initial five-year period. Between the 5th and the 10th years of the term an area of not less than 210,000 acres shall be developed for the purpose of ranching domestic livestock, and a minimum of 8,000 head of cattle, or their equivalent,

......./4

Ref: 65678/IV/72   7th June, 1967

- 4 -

shall be introduced into the area. By the 15th year of
the term, a total area of not less than 420,000 acres
shall have been developed for the purposes of ranching
domestic livestock, and a minimum of 17,000 head of
cattle, or their equivalent, shall be introduced into
the area. By the 20th year of the term a minimum area
of 640,000 acres shall have been developed for the
purpose of ranching domestic livestock, and a minimum
of 26,000 head of cattle, or their equivalent, shall
have been introduced into the area. The Grantee shall
be required to carry out all forms of ancillary
development, including bush clearance, pasture improvement, development of water resources, provisions of
means of access, construction of requisite paddocks and
all other arrangements necessary for the proper pasture
and control of livestock in the numbers stated.

5. The Grantee shall so manage the natural animal population
of the game and tourist development area that at no time,
due to the Grantee's operations, shall the area become
denuded of such animal population, and the Grantee's
management of the area shall be so conducted as to ensure
that the overall animal population is encouraged to
increase within the capacity of the area to accept such
increase.

6. The Grantee shall throughout the term maintain the
improvements made on the land in good and substantial
repair and condition.

7. The Grantee shall submit to the Commissioner of Lands
in each year for the first 20 years of the term, a
detailed report of the improvements affected on the
land within the preceding year, including details of all
wild animals killed or captured and trophies found and
the steps being taken for the eradication of tsetse-fly
and the introduction of cattle and other livestock into
the area.

8. The Grantee shall produce to the Commissioner of Lands
audited accounts complying with the provisions of the
Companies' Act (CAP. 486) not later than 31st March of
each year following the year in which the operations
to which the accounts relate, were carried out.

9. The Grantee shall pay such rates, taxes, charges, duties,
assessments or outgoings of whatever description as may
be levied by the Government or Local Authority upon the
land or the buildings erected thereon in accordance with
the scales levied by the Government or Local Authority
in respect of similar land elsewhere.

10. The Grantee shall not subdivide, transfer, charge,
sublet or otherwise part with the possession of the
land or any part thereof without the prior written
consent of the President. Providing that no such consent
shall be required in respect of the subletting of residential
accommodation in lodges or hotels.

11. The Grantee shall not erect any permanent buildings on
the land or make external alterations to such permanent
buildings otherwise than in conformity with plans and
specifications previously approved in writing by the
Commissioner of Lands and the Local Authority.

......../5

Ref: 65678/IV/7⚫️   7th June , 1967.

- 5 -

12.   The Grantee shall pay to the Ministry of Tourism and
       Wildlife a sum to be agreed by negotiation for the
       existing buildings situated on the north bank of the
       Galana River in the vicinity of Lali Hills.

13.   The President or such authority or person as may be
       appointed for the purpose shall have the right to enter
       upon the land and lay and have access to water mains,
       service pipes and drains, telephone or telegraph wires
       and electric mains of all descriptions whether overhead
       or underground and the Grantee shall not erect any
       building in such a way as to cover or interfere with any
       existing alignments of main or service pipes or telephone
       or telegraph wires and electric mains.

14.   The Grantee shall be permitted to carry out Professional
       Hunting in accordance with the provisions of the Wild
       Animals Protection Act (CAP. 376), providing that :-

       (a) Hunters shall pay the licence fees required by
            the Game Department subject to Special Conditions
            14(c) and 16(d).

       (b) Overseas Visitors must be assisted by the holder
            of an Assistant's Permit.

       (c) Overseas Visitors shall be permitted to participate
            in game cropping.

       (d) All game animals shot on the Grantee's land shall
            be deducted from the animal quota allocated by the
            Game Department, in accordance with Special Condition
            16(b).

15.   The Grantee shall only conduct the trapping and export of
       wild animals in accordance with the recommendations of
       the Advisory Committee on the Capture and Export of Live
       Animals.

16.   Game cropping shall be carried out by the Grantee in
       accordance with the following rules :-

       (a) The Grantee may utilize the products of cropped
            animals.

       (b) Game cropping shall be in accordance with an
            annual quota allotted by the Game Department and the
            Grantee shall rigidly adhere to the terms of the
            allotment of the quota. Both maximum and minimum
            quotas will be stipulated.

       (c) Hunting for cropping purposes shall only be
            carried out by persons approved by the Game
            Department.

       (d) The Grantee shall pay to the Government of Kenya
            K.Shs. 100/- in respect of each elephant killed in
            the process of game cropping.

17.   The Grantee shall at any time permit the President, or such
       authority or person as may be appointed for the purpose, to
       enter upon the land to carry out wild life and range
       research projects, including such projects which necessitate
       the killing of game.

18.   All game trophies found on the land shall remain the
       property of the Government of Kenya.

                                              ..../6

Ref: 65678/IV/72   7th June , 1967.

- 6 -

19. The Grantee shall be required to retain the services
     of a minimum of five game scouts, and shall offer
     employment to other members of the existing labour
     force in the area on a basis to be agreed with the Kenya
     Government.

20. The Grantee shall carry out water development in the area
     between the northern boundary of the Scheme and the Tiva
     River in accordance with the wishes of the Orma people
     and to the following minimum requirements :-

     (a) In the first year of the term to install one hafia
          dam of not less than 1 million gallons capacity at
          a site to be selected in consultation with the Orma
          and the Government.

     (b) Within two years of the commencement of the term
          to install a second hafia dam of not less than
          1 million gallons capacity at a site similarly
          selected.

     (c) In the third and subsequent years of the term to
          either establish further hafia dams at the rate of
          one each year until such time as the Orma have full
          access to the grazing between the Tiva River and the
          northern boundary of the Scheme or if such dams are
          not producing satisfactory results, make permanent
          water installations on the Tiva River to the west of
          Assa.

21. Notwithstanding anything contained herein or in the said
     Government Lands Act, the Grantee shall, on receipt of six
     months' notice in writing in that behalf surrender all or
     part of the land which shall be required for construction
     of roads, railways, Government installations, schools,
     hospitals, dispensaries, military installations, landing
     grounds, or such like purpose as may, at any time, be
     specified. In the event of any such surrender being
     required as aforesaid, the Company shall be paid fair and
     reasonable compensation for any permanent improvements
     affected on the land, such compensation to be assessed by
     the Commissioner of Lands, but no compensation shall be
     payable in respect of severance of part of the land by
     reason of such surrender.

                    *   *   *   *

     If acceptance of the above-mentioned Letter of Allotment,
and payment of the annual sum of K.Shs. 3,000/= in respect of the
period July to December 1967, are not received within thirty days
from the date hereunder, this offer will be considered to have
elapsed.

                         I have the honour to be,
                              Gentlemen,
                         Your obedient servant,

                              (J.A. O'Loughlin)
                              COMMISSIONER OF LANDS

                                             ..../7

Ref: 65678/IV/72   7th June , 1967.

- 7 -

Copy to :-

The Permanent Secretary,
Ministry of Tourism & Wildlife,
NAIROBI

The Permanent Secretary,
Ministry of Agriculture & Animal Husbandry,
NAIROBI

The Permanent Secretary,
The Treasury,
NAIROBI

The Provincial Commissioner,
Coast Province,
MOMBASA

The Director of Surveys,
NAIROBI

The Provincial Agricultural Officer,
MOMBASA

The District Commissioner,
Tana River District,
P.O. GALOLE via MALINDI

Messrs. Daly & Figgis,
Advocates,
P.O. Box 34,
NAIROBI

S.P.R.O.   )
Land Rents   )   To Note Please.
O/C Records   )

WILDLIFE POPULATION ANALYSIS

Direr· Anderson, Chairman (U.S.) J.G.L. Powys, M.G. Prettejohn, The Hon. R. Matano, A.D.G. Dyer, A.S. Atherton (U.S.) L.J. Ayuko

GALANA SAFARIS
GALANA GAME & RANCHING LTD.
(HUNTING AND PHOTOGRAPHIC SAFARIS)
MEMBER: EAST AFRICAN PROFESSIONAL HUNTERS ASSOCIATION
INTERNATIONAL PROFESSIONAL HUNTERS ASSOCIATION

United States Contact:-
M. Anderson
415/393 — 2055 (Office Hours)
415/4410216 (After Office Hours)
San Francisco, U.S.A.

All Correspondence & Enquiries should be referred to:
17th. June, 1977.

P.O. Box 76 Malindi, Kenya.
Radiocall: 2007
Telegrams: "LALI" Malindi, Kenya.
Emergency Telephone No.
25853 Nairobi (Office Hours)

WILDLIFE POPULATION ANALYSIS.

The Company, voluntarily and at regular intervals carries out
wildlife monitoring. The Cpmpany aircraft is used extensively
to assist with these game counts. The following list gives an
indication of wildlife increases by species:-

SPECIES	1962	1968	1972	1977
Elephant	1,851	1,430	2,166	5,495*
Rhinoceros	unknown	192	134	98*
Buffalo	"	2,020	3,444	4,432*
Lion	"	unknown	207	235
Leopard	"	"	69	50
Eland	"	391	1,300	2,378*
Giraffe	"	640	855	1,036*
Common Zebra	"	1,320	1,603	2,200*
Oryx Fringe Eared	"	4,380	5,279	8,743*
Waterbuck Common	"	295	321	434
Hartebeeste Cokes	"	16	83	177*
Topi	"	18	35	42
Lesser Kudu	"	unknown	497	364
Gerenuk	"	"	790	840
Peters Gazelle	"	"	2,608	2,800*
Impala	"	90	138	227
Bushbuck	"	unknown	unknown	60
Oribi	"	"	"	42
Duiker Grey	"	"	320	368
Warthog	"	"	588	689
DikDik	"	"	1,320	1,586
Cheetah	"	"	29	137
Hyena	"	"	288	304
Jackal	"	"	69	127
Crocodile	"	150	unknown	270
Ostrich	"	unknown	743	1,091
Guinea fowl	"	"	8,924	11,260
Sandgrouse			several thousands	

Species marked with * indicate that the figures are taken from an
average of two different aerial game counts. The two counts were
carried out on the 18th. and 19th. October, 1976 and the 7th. and
8th. April, 1977 respectively. The Elephant count in 1962 and 1968
was an aerial count carried out on this property by the Tsavo Park
authorities. All other counts were carried out by ground
observation, using cónstant routes and strips. Apart from three
species, it appears that all the wildlife on this property is
increasing most remarkably. The exceptions are, Lesser Kudu,
which have been drastically reduced by a severe rinderpest
infestation. The Rhinoceros and Leopard populations have been
reduced by poaching activity.

The unusual and rapid increase in the resident Elephant population

CONT/..........

- 2 -

is thought to have been caused by the denudation of Elephant
habitat in the adjoining Tsavo Park(East), and also to the
continuous poaching activities that constantly harrass the
Elephant herds to the North and South of this property.   In
contrast, this property can offer adequate vegetation, also
fairly good protection from poachers is acheived, because of the
constant operation of the resident mobile Anti-poaching Unit.

K. Clark

KEN CLARK.
GAME MANAGER.